Studies in Computational Intelligence

Volume 661

Series editor

Janusz Kacprzyk, Polish Academy of Sciences, Warsaw, Poland
e-mail: kacprzyk@ibspan.waw.pl

About this Series

The series "Studies in Computational Intelligence" (SCI) publishes new developments and advances in the various areas of computational intelligence—quickly and with a high quality. The intent is to cover the theory, applications, and design methods of computational intelligence, as embedded in the fields of engineering, computer science, physics and life sciences, as well as the methodologies behind them. The series contains monographs, lecture notes and edited volumes in computational intelligence spanning the areas of neural networks, connectionist systems, genetic algorithms, evolutionary computation, artificial intelligence, cellular automata, self-organizing systems, soft computing, fuzzy systems, and hybrid intelligent systems. Of particular value to both the contributors and the readership are the short publication timeframe and the worldwide distribution, which enable both wide and rapid dissemination of research output.

More information about this series at http://www.springer.com/series/7092

Iuliana F. Iatan

Issues in the Use of Neural Networks in Information Retrieval

Iuliana F. Iatan
Department of Mathematics and Computer
Sciences
Technical University of Civil Engineering
Bucharest
Romania

ISSN 1860-949X ISSN 1860-9503 (electronic)
Studies in Computational Intelligence
ISBN 978-3-319-82930-2 ISBN 978-3-319-43871-9 (eBook)
DOI 10.1007/978-3-319-43871-9

Printed on acid-free paper

This Springer imprint is published by Springer Nature
The registered company is Springer International Publishing AG
The registered company address is: Gewerbestrasse 11, 6330 Cham, Switzerland

The measure of success for a person is the magnitude of his/her ability to convert negative conditions to positive ones and achieve goals.

—G.A. Anastassiou

Contents

Introduction

The pattern recognition field is a fundamental one and it is in a full expansion direction of the information technology. The pattern search in the data has a long and a full success history. The observation of some regular patterns in the planet motion has determined Kepler to the discovery of the empirical laws of the planetary motion. The discovery of some regularities in the atomic spectrum has played a key role in the quantum physical development.

In the last years, pattern recognition (PR) has some essential applications in the biometry, satellite image analysis for the detection and the assessment of the terrestrial resources, robotics, medicine, biology, psychology, marketing, computer vision, artificial intelligence, and remote sensing. Other effervescent fields have detached from the PR field: data mining, Web searching, retrieval of multimedia data. Recently, "a lot of area comes under Pattern Recognition due to emerging application which are not only challenging but also computationally more demanding"[1] (see Fig. 1).

"Pattern recognition is not only about methods; it is about taking a new view of the problem at hand that allows one to single out the principal point of interest in a large volume of information and suggest a nontrivial solution."[2]

The statistical approach has been most intensively studied and used in practice than the traditional approaches of pattern recognition. However, the theory of artificial neural network techniques has been getting significant importance. "The design of a recognition system requires careful attention to the following issues:

---

[1]Dutt, V., and Chadhury, V., and Khan, I., Different Approaches in Pattern Recognition, Computer Science and Engineering, 2011, 1(2), 32–35.

[2]Neimark, Yu.I., and Teklina, L.G., On Possibilities of Using Pattern Recognition Methods to Study Mathematical Models, Pattern Recognition and Image Analysis, 2012, 22(1), 144–149.

Problem Domain	Application	Input Pattern	Pattern classes
Bioinformatics	Sequence Analysis	DNA/ Protein Sequence	Known types of genes patterns
Data Mining	Searching for meaningful patterns	Points in multi- dimensional space	Compact and well separated clusters
Document image analysis	Reading machine for the blind	Document image	Alphanumeric characters, words
Multimedia database retrieval	Internet search	Video clip	Video genres(e.g. action, dialogue,etc.)
Biometric recognition	Forcasting crop yield	Multispectral image	Land use categories, growth pattern of crops
Speech recognition	Telephone directory enquiry without operator assistance	Speech waveform	Spoken words

Fig. 1 Example of pattern recognition applications

definition of pattern classes, sensing environment, pattern representation, feature extraction and selection, cluster analysis, classifier design and learning, selection of training and test samples, and performance evaluation."[3]

Based on presence or absence of teacher and the information provided for the system to learn there are three basic types of learning methods in neural networks [1]:

1. Supervised learning;
2. Unsupervised learning;
3. Reinforced learning;

These are further categorized, based on the rules used, as follows:

- Hebbian,
- Gradient descent,
- Competitive,
- Stochastic learning.

[3]Basu, J.K., and Bhattacharyya, D., and Kim, T.H., Use of Artificial Neural Network in Pattern Recognition, International Journal of Software Engineering and Its Applications, 2010, 4(2), 22–34.

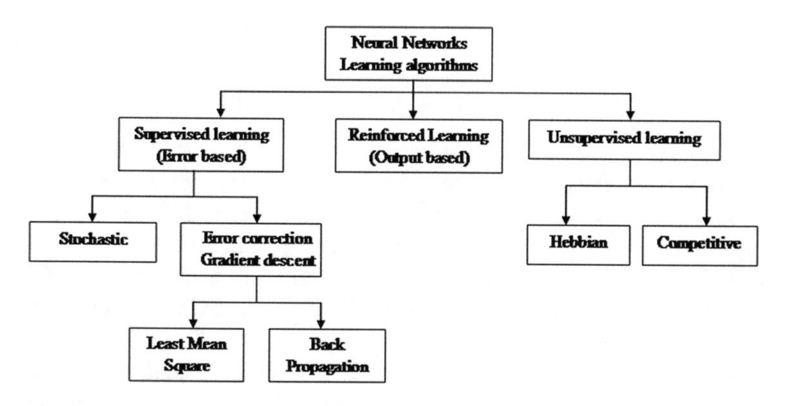

Fig. 2 Classification of learning algorithms

Figure 2 indicates the hierarchical representation of the previous mentioned algorithms.

The four best known approaches for pattern recognition are as follows:

1. Statistical classification
2. Syntactic or structural matching
3. The recognition with intelligence computational techniques
4. The recognition with the expert systems based on some languages specific to the artificial intelligence.

Neural computing is an information processing paradigm, inspired by biological system, composed [2] of a large number of highly interconnected processing elements (neurons) working in unison to solve specific problems.

Neural network constitutes an important component in Artificial Intelligence (AI). "It has been studied for many years in the hope of achieving human-like performance in many fields, such as classification, clustering, and pattern recognition, speech and image recognition as well as information retrieval by modeling the human neural system."[4]

Artificial neural network (ANN) represents an adaptive mathematical model or a computational structure, designed to simulate a system of biological neurons, which transfers an information from its input to output in a desired way. An artificial neural network has a lot of interconnecting artificial neurons to employ some mathematical or computational models for information processing. Among the advantages of the neural networks are: learning, adaption, fault tolerance, parallelism, and generalization.

For several years now, neural networks (NNs) have been applied in many different application domains in order to solve various information processing

[4]Reshadat, V., and Feizi-Derakhshi, M.R., Neural Network-Based Methods in Information Retrieval, American Journal of Scientific Research, 2011, 58, 33–43.

problems of regression, classification, computational science, computer vision, data processing, and time series analysis. Hence, they have enjoyed wide popularity [3].

Neural network applications can be grouped [1] in the following categories:

(A) **Clustering**: It is a "method to organize automatically a large data collection by partition a set data, so the objects in the same cluster are more similar to one another than with the objects belonging to other clusters."[5] The representative applications include data compression and data mining.
(B) **Classification/Pattern Recognition**: It has the task to assign an input pattern to one of many classes. This category includes algorithmic implementations such as associative memory.
(C) **Function Approximation**: As its aim is to find an estimate of the unknown function subject to noise, it is required by various engineering and scientific disciplines.
(D) **Prediction Systems**: They have the task to forecast some future values of a time-sequenced data. Prediction is unlike the function approximation by considering time factor. Hence, the dynamic system may produce different results for the same input data based on time, which means the system state.

Information retrieval (IR) represents a wide research area mainly on the Internet. IR "is concerned with the analysis, representation and retrieval of texts."[6]

IR "is different from data retrieval in databases using SQL queries because the data in databases are highly structured and stored in relational tables, while information in text is unstructured. There is no structured query language like SQL for text retrieval."[7]

Data mining constitutes another layer of data processing, introduced in order to have a better perception of information for management and decision making. The main aim of this processing layer is [4] to:

(i) extract the implicit, hidden, potentially useful information;
(ii) discover meaning full patterns from large raw data collections.

Data mining is "a multidisciplinary field involving machine learning, statistics, databases, artificial intelligence, information retrieval, and visualization." (see footnote 7).

Some of the common data mining tasks are [5]: supervised learning (or classification), unsupervised learning (or clustering), association rule mining, sequential pattern mining, and regression.

Document clustering is a fundamental task in text mining, being concerned with grouping documents into clusters according to their similarity; more exactly,

---

[5]Bharathi, G. and Venkatesan, D., Improving Information Retrieval using Document Clusters and Semantic Synonym Extraction, Journal of Theoretical and Applied Information Technology, 2012, 36(2), 167–173.

[6]Reshadat, V., and Feizi-Derakhshi, M.R., Neural Network-Based Methods in Information Retrieval, American Journal of Scientific Research, 2011, 58, 33–43.

[7]Liu, B., Web DataMining, Springer-Verlag Berlin Heidelberg, 2008.

document clustering achieves automatically group of the documents that belong to the same topic, in order to provide user's browsing of retrieval results. "Document clustering has always been used as a tool to improve the performance of retrieval and navigating large data."[8] The representative applications include data compression and data mining.

Being suited for information retrieval from large text to multimedia databases, the NNs have been widely used in the area of IR and text mining, such as text classification, text clustering, and collaborative filtering. "In recent years, with the fast growth of the World Wide Web and the Internet, these algorithms have also been used in Web-related applications such as Web searching, Web page clustering, Web mining, etc. Their capacity for tolerant and intuitive processing offers new perspectives in information retrieval."(see footnote 6).

When large volumes of data are to be handled, a suitable approach to increase the IR speed is to use the NNs as an artificial intelligent technique. I was motivated to write this book to highlight the ability of the NNs to be very good pattern matchers and the importance of the NNs for the IR, which is based on index term matching.

Chapter 1 of this book emphasizes the possibility of using neural networks in IR and highlights the advantages of applying two neural networks models for solving the problem of simplifying the complex structure of an IR system, by substitution of the relations between its subsystems by NNs.

The core to measure similarity or distance between two information entities is required for all information discovery tasks (whether IR or data mining). It is crucial to use an appropriate measure both to improve the quality of information selection and to reduce the time and processing costs [4]. The relevance of the concept of similarity has proven [4] not only in every scientific field but in philosophy and psychology, too. However, this work deals with the measure of similarity in computer science domain (information retrieval and data mining to be more specific). In the thesis domain, the similarity measure "is an algorithm that determines the degree of agreement between entities."[9]

"Similarity-based classifiers estimate the class label of a test sample based on the similarities between the test sample and a set of labeled training samples, and the pairwise similarities between the training samples."[10]

Image similarity is an important concept in many applications. In Chap. 2 of this book we define a new neural network based method for learning image similarity [6]. We start off from the fuzzy Kwan-Cai neural network [7] and turn it into a partially supervised one. Using the training algorithm of the FKCNN, its third and

---

[8]Bharathi, G. and Venkatesan, D., Improving Information Retrieval using Document Clusters and Semantic Synonym Extraction, Journal of Theoretical and Applied Information Technology, 2012, 36(2), 167–173.

[9]Zaka, B., Theory and Applications of Similarity Detection Techniques, 2009, http://www.iicm.tugraz.at/thesis/bilal_dissertation.pdf.

[10]Chen, Y., and Garcia, E.K., and Gupta, M.Y., and Rahimi, A., and Cazzanti, A., Similarity-based Classification: Concepts and Algorithms, Journal of Machine Learning Research, 2009, 10, 747–776.

fourth layers are built during the learning process. The training stage is followed by a calibration stage, in which the fuzzy neurons (FNs) of the fourth layer will be assigned category labels. We performed a comparative studiy of the proposed similarity learning method and compared it to self-organizing Kohonen maps (SOKM) and *k*-nearest neighbor rule (*k*-NN). The resulting similarity functions are tested on the VOC data set that consists in 20 object classes. The results indicate that the neural methods FKCNN and SOKM are performing better for our task than *k*-NN. SOKM sometimes gives good results, but this depends highly on the right parameter settings. Small variations induced large drops in performance. The overall performance of FKCNN is better. The main advantage of FKCNN consists in the fact that we can obtain good results that are robust to changes in the parameter settings.

There is a section in Chap. 2, where we have performed the software implementation of the FKCNN to be experimented for the face recognition task, using the ORL Database of Faces, provided by the AT&T Laboratories from Cambridge University; it contains 400 images, corresponding to 40 subjects (namely, 10 images for each of the 40 classes).

Recently, ANNs methods have become useful for a wide variety of applications across a lot of disciplines and in particularly for prediction, where highly nonlinear approaches are required [8]. The advantage of neural networks consists [9] in their ability to represent both linear and nonlinear relationships and to learn these relationships directly from the data being modeled. Among the statistical techniques that are widely used is the regression method, the multiple regression analysis has the objective to use independent variables whose values are known to predict the single dependent variable.

Our research objective in Chap. 3 is to compare [10] the predictive ability of multiple regression and fuzzy neural model, by which a user's personality can be accurately predicted through the publicly available information on their Facebook profile. We shall choose to use the fuzzy Gaussian neural network (FGNN) for predicting personality because it handles nonlinearity associated with the data well.

Function approximation has the aim to find the underlying relationship from a given finite input–output data, being a fundamental problem in a lot of real-world applications, such as prediction, pattern recognition, data mining, and classification. Various methods have been developed to solve the problem of function approximation, one of them being with artificial neural networks [11]. The problem of estimating a function from a set of samples means an advance in the field of neural networks, as there has been much research work being carried out in exploring the function approximation capabilities of NN's [12]. "Approximation or representation capabilities of neural networks and fuzzy systems have attracted considerable research in the last 15 years."[11]

[11]Zeng, X.J., and Keane, J.A., and Goulermas, J.Y., and Liatsis, P., Approximation Capabilities of Hierarchical Neural-Fuzzy Systems for Function Approximation on Discrete Spaces, International Journal of Computational Intelligence Research, 2005, 1, 29–41.

The Fourier series neural networks (FSNNs) represent one type of orthogonal neural networks (ONN) and they are feedforward networks, similar to sigmoidal neural networks. After we have studied the FSNN for function approximation in Chap. 4, we have designed a new neural model [13]. FSNN performs only to approximate trigonometric functions and not for all the kind of functions. Therefore, it was necessary to built in the paper [13] a four layer neural network which works very well both for the approximation of the trigonometric functions and for other types of functions (as multivariate polynomial, exponential) too. The main advantage of our proposed model consists in its special weight matrix. This matrix has different expressions in terms of the function that has to be approximated.

For each function, the approximation achieved using our neural method is finer than that obtained using the FSNN. Approximating a function with an FSNN is better than using a Tayor series as it has a smaller maximum error and is more economical. There are many advantages of using neural network to implement function approximation.

The field studied in this chapter has significant references for the research of function approximation and all the conclusions are proved effective after the actual simulation tests.

The idea of Chap. 5 is to develop the neural networks in other than the real domain. "Neural computation in Clifford algebras, which include familiar complex numbers and quaternions as special cases, has recently become an active research field."[12] It is interesting to use complex numbers in the context of neural networks as they tend to improve learning ability [14]. "Though neurons with plural real or complex numbers may be used to represent multidimensional data in neural networks, the direct encoding in terms of hyper-complex numbers may be more efficient."[13]

In order to achieve the aim of illustrating the usefulness of the Clifford algebra in the neural computing because of its geometric properties, we have introduced in Chap. 5, the Fuzzy Clifford Gaussian network (FCGNN), contributing [15] in this way to continue the development of neural networks in other than the real domain.

Chapter 6 introduces two concurrent neural models:

1. *Concurrent fuzzy nonlinear perceptron modules* (CFNPM), representing a winner-takes-all collection of small FNP (Fuzzy Nonlinear Perceptron) units;
2. *Concurrent fuzzy Gaussian neural network modules* (CFGNNM), which consists of a set of M fuzzy neural networks, by the type FGNN.

[12]Buchholz, S., and Sommer, G., Introduction to Neural Computation in Clifford Algebra, 2010, http://www.informatik.uni-kiel.de/inf/Sommer/doc/Publications/geocom/buchholz_sommer1.pdf.

[13]Matsui, N., and Isokawa, T., and Kusamichi, H., and Peper, F., and Nishimura, H., Quaternion neural network with geometrical operators, Journal of Intelligent & Fuzzy Systems, 2004, 15, 149–164.

The use of them for face recognition causes an increase in the recognition rates for the training and the test lot (both in the case of making the feature selection), compared to those achieved using the simple variations of FNP and FGNN.

The aim of Chap. 7 is to design a new model of fuzzy nonlinear perceptron, based on alpha level sets, entitled as fuzzy nonlinear perceptron based on alpha level sets (FNPALS). It differs from the other fuzzy variants of the nonlinear perceptron, where the fuzzy numbers are represented by membership values, i.e., in the case of the FNPALS, the fuzzy numbers are represented through the alpha level sets.

The last chapter has the aim to describe a recurrent fuzzy neural network (RFNN) model, whose learning algorithm is based on the improved particle swarm optimization (IPSO) method. The proposed RFNN is different from other variants of RFNN models through the number of the evolution directions that they use: we update the velocity and the position of all particles along three dimensions [16], while in [17] two dimensions are used.

References

1. R.C. Chakraborty. Fundamentals of neural networks. http://www.myreaders.info/html/artificial_intelligence.htm, 2010.
2. J.K. Basu, D. Bhattacharyya, and T.H. Kim. Use of artificial neural network in pattern recognition. *International Journal of Software Engineering and Its Applications*, 4(2):22–34, 2010.
3. L.A. Gougam, M. Tribeche, and F. Mekideche-Chafa. A systematic investigation of a neural network for function approximation. *Neural Networks*, 21:1311–1317, 2008.
4. B. Zaka. Theory and applications of similarity detection techniques. http://www.iicm.tugraz.at/thesis/bilal_dissertation.pdf, 2009.
5. B. Liu. *Web DataMining*. Springer-Verlag Berlin Heidelberg, 2008.
6. I. Iatan and M. Worring. A fuzzy Kwan- Cai neural network for determining image similarity. *BioSystems (Under Review)*, 2016.
7. H.K. Kwan and Y. Cai. A fuzzy neural network and its application to pattern recognition. *IEEE Trans. on Fuzzy Systems*, 2(3):185–193, 1997.
8. V. Bourdès, S. Bonnevay, P. Lisboa, R. Defrance, D. Pérol, S. Chabaud, T. Bachelot, T. Gargi, and S. Nègrier. Comparison of artificial neural network with logistic regression as classification models for variable selection for prediction of breast cancer patient outcomes. *Advances in Artificial Neural Systems*, pages 1–10, 2010.
9. O.S. Maliki, A.O. Agbo, A.O. Maliki, L.M. Ibeh, and C.O. Agwu. Comparison of regression model and artificial neural network model for the prediction of electrical power generated in Nigeria. *Advances in Applied Science Research*, 2(5):329–339, 2011.
10. I. Iatan and M. de Rijke. Predicting human personality from social media using a fuzzy neural network. *Neural Computing and Applications (Under Review)*, 2016.
11. Z. Zainuddin and P. Ong. Function approximation using artificial neural networks. *WSEAS Transactions on Mathematics*, 7(6):333–338, 2008.
12. S.S. Haider and X.J. Zeng. Simplified neural networks algorithm for function approximationon discrete input spaces in high dimension-limited sample applications. *Neurocomputing*, 72:1078–1083, 2009.
13. I. Iatan, S. Migorski, and M. de Rijke. Modern neural methods for function approximation. *Applied Soft Computing*, (Under Review), 2016.

14. N. Matsui, T. Isokawa, H. Kusamichi, F. Peper, and H. Nishimura. Quaternion neural network with geometrical operators. *Journal of Intelligent & Fuzzy Systems*, 15:149–164, 2004.
15. I. Iatan and M. de Rijke. A fuzzy Gaussian Clifford neural network. (work in progress), 2014.
16. G.A. Anastassiou and I. Iatan. A recurrent neural fuzzy network. *Journal of Computational Analysis and Applications*, 20 (2), 2016.
17. C.J. Lin, M. Wang, and C.Y. Lee. Pattern recognition using neural-fuzzy networks based on improved particle swam optimization. *Expert Systems with Applications*, 36:5402–5410, 2009.

Chapter 1
Mathematical Aspects of Using Neural Approaches for Information Retrieval

Scientists have shown considerable interest in the study of Artificial Neural Networks (NNs) during the last decade. Interest in Fuzzy Neural Network (FNN) applications was generated [1] by two events:

1. first, the success of Japanese fuzzy logic technology applications in consumer products;
2. second, in some neural network applications sufficient data for training are not available.

In such situations, fuzzy logic systems are often workable.

Today, various neural models are established [2] as fixed parts of machine learning, and a thorough theoretical investigation for these models is available.

A lot of researchers of this scientific topic agree on the fact that the statistical notion [3] is often "the right language to formalize learning algorithms and to investigate their mathematical properties."[1]

Nevertheless, according to the widespread models, tasks, and application fields of the NNs, "mathematical tools ranging from approximation theory, complexity theory, geometry, statistical physics, statistics, linear and nonlinear optimization, control theory and many more areas can be found in the neural literature." (see footnote 1).

Correspondingly, the role of mathematics in the neural network literature is [2] diverse:

(A) **development and presentation of algorithms**: Most neural algorithms are formulated in mathematical terms and some learning schemes are even mainly inspired by some abstract mathematical considerations such as support vector machines;

[1]Hammer, B., and Villmann, T., Mathematical Aspects of Neural Networks, 11th European Symposium on Artificial Neural Networks (ESANN' 2003), 2003, 59–72.

I.F. Iatan, *Issues in the Use of Neural Networks in Information Retrieval*,
Studies in Computational Intelligence 661, DOI 10.1007/978-3-319-43871-9_1

(B) **foundation of tools**: A fixed canon of mathematical questions has been recognized for most network models and application fields which is to be answered in the aim of establishing the models as well founded and strong tools in the literature. "Interestingly, many mathematical questions are thereby not yet solved satisfactorily also for old network models and constitute still open topics of ongoing research such as the loading problem of feedforward networks or the convergence problem of the self-organizing map in its original formulation"[2];

(C) **application of tools**: Mathematical formalization finds templates for the assessment of the performance of methods and application in real-life scenarios (although these templates are not always followed and real life would sometimes be better described by slightly different features than the Mathematics).

We shall in the following assume mathematical questions, "which are to be answered to justify standard models as reliable machine learning tools." (see footnote 2).

Thereby, we shall focus on the classical models used for machine learning, namely on the feedforward networks.

The goal of using the feedforward networks depends [2] on the output function of such type networks; they can be used for the:

- classification of patterns if the output set is finite and discrete;
- approximation of functions if the output set is contained in a real vector space.

"The development of Information Technology has generated large amount of databases and huge data in various areas. The research in databases and information technology has given rise to an approach to store and manipulate this precious data for further decision-making."[3]

Data mining is an interdisciplinary field, the confluence of a set of disciplines, including database systems, statistics, machine learning, visualization, and information science. Moreover, depending on the data mining approach used, techniques from other disciplines may be applied, such as neural networks, fuzzy and/or rough set theory, knowledge representation, inductive logic programming, or high performance computing. Depending on the kinds of data to be mined or on the given data mining application, the data mining system may also integrate techniques from spatial data analysis, information retrieval, pattern recognition, image analysis, signal processing, computer graphics, Web technology, economics, or psychology.

Information Retrieval System (IRS) represents [4] a system designed to store items of information that need to be processed, searched, and retrieved according to a user's query. The IRSs are nowadays very popular mainly due to the popularity of the Web. The many text pages in the World Wide Web [5] determine the results that search engines provide to contain a lot of irrelevant information. Therefore, finding the information needed by a user is a difficult and complicated task and it has increased the importance of proper retrieval of information.

[2]Hammer, B., and Villmann, T., Mathematical Aspects of Neural Networks, 11th European Symposium on Artificial Neural Networks (ESANN' 2003), 2003, 59–72.

[3]Ramageri, B.M., Data Mining Techniques and Applications, Indian Journal of Computer Science and Engineering, 2010, 1(4), 301–305.

"As the amount of documents become more and more higher, the chance to find the proper information is more and more lower. Therefore, retrieving the proper information in little time is a necessity. The Internet requires new techniques, or extensions to existing methods, to address gathering information, making index structures scalable and efficiently updatable. Therefore, it can improve the ability of search engines to discriminate useful from useless information."[4]

1.1 Information Retrieval Models

Web classification has been studied through a lot of different technologies. The crucial used models in IRSs and on the Web are [6]:

(1) **Set-theoretic models**
The information retrieval models based on set theory are:

- *Boolean model* is the simplest form of an IR model based on typical operations AND, OR and NOT, which is referred to as exact match model. The simple Boolean retrieval model has the advantage of being easy to implement and computationally efficient. Its drawbacks consist of: complex queries are hard to construct, unavailability of ranked results, and no partial matching (rather extreme output of either logical match or no match).
- *fuzzy set based model* is practiced in IRSs in order to address the shortcoming of simple Boolean model with strict binary association. As the information entities are assigned to a degree of membership, the model introduces the membership/association, a gradual notion rather than binary.
- *extended Boolean model* adds value to simpler model through the ability of weight assignment, and use of positional information. The term weights added to data objects help generate the ranked output. It is a combination of vector model characteristics and Boolean algebra.

(2) **Algebraic models**
Besides the logical reasoning approach IR models, there are others based on algebraic calculus, like:

- *Vector Space Model (VSM)* (see Fig. 1.1) is used to represent the information as vectors in multidimensional space. Each dimension corresponds to a possible feature of the information (e.g., term in document). A distance function applied to the information vectors provides the match and rank information. VSM-based information retrieval is considered a good mathematical implementation for processing large information sources. It provides possibilities of partial matching and ranked result output. However, this approach lacks the

[4]Reshadat, V., and Feizi-Derakhshi, M.R., Neural Network-Based Methods in Information Retrieval, American Journal of Scientific Research, 2011, 58, 33–43.

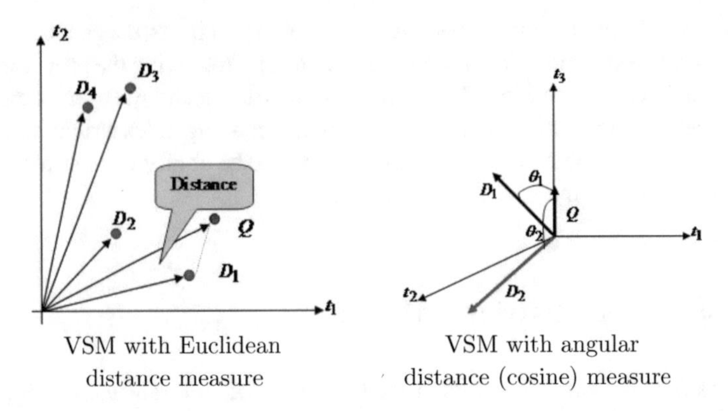

Fig. 1.1 Vector space model

control of Boolean model, and has no means to handle semantic or syntactical information.

- *Latent Semantic Analysis (LSA) based model* converts the large information matrix (term-document) to a lower dimensional space using singular value decomposition (SVD) technique.
- *Neural Networks* uses the weighted and interconnected representation of information.

(3) **Probabilistic models**
The models based on probabilistic inferences include:

- *inference network* (represented in Fig. 1.2) consists of a document network and a query network.
- *belief network* IR model is a generalized form of inference network model having a clearly defined sample space.

(4) **Knowledge-based models**
The knowledge-based IR models use formalized linguistic information, structural and domain knowledge to discover semantic relevance between objects.

(5) **Structure-based models**
The structural information retrieval models combine the content and structural characteristics to achieve greater retrieval efficiencies in many applications.

A lot of algorithms and techniques like Classification, Clustering, Regression, Artificial Intelligence, Neural Networks (NNs), Association Rules, Decision Trees, Genetic Algorithm, Nearest Neighbor method, etc., are attempted [7] for knowledge discovery from databases.

Because of increasing the volume of information, the deficiency of traditional algorithms for fast information retrieval is [5] more evident. When large volumes of data are to be handled, the use of NNs as an Artificial Intelligent technique is suitable approach to increase the Information Retrieval (IR) speed. The NNs can

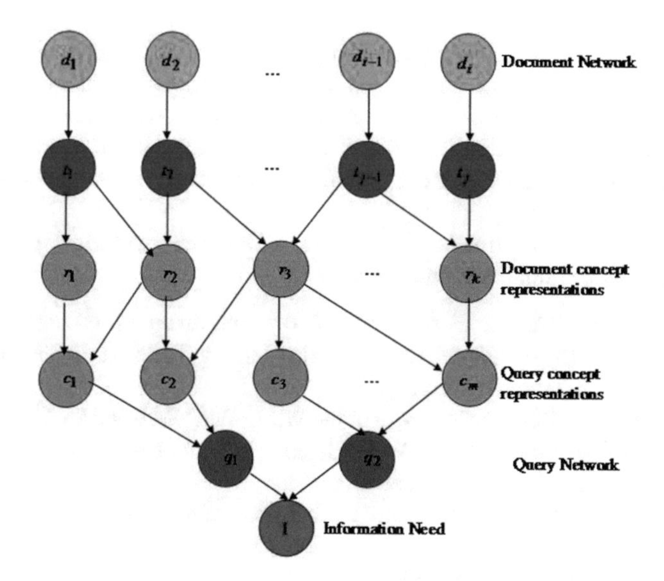

Fig. 1.2 Document inference network

reduce "the dimension of the document search space with preserving the highest retrieval accuracy."[5]

Neural network can be considered one of the important components in Artificial Intelligence (AI) as it has been studied [5] for many years in the hope of achieving human-like performance in many fields, like classification, clustering, and pattern recognition, speech and image recognition as well as information retrieval by modeling the human neural system.

A NN represents "an oversimplified representation of the neuron interconnections in the human brain,"[6] in the sense that [8]:

- the nodes constitute the processing units;
- the arcs (edges) mean the synaptic connections;
- the strength of a propagation signal can be simulated by a weight, which is associated to each edge;
- the state of a neuron is defined by its activation function;
- the axon can be modeled by the output signal which is issued by a neuron depending on its activation function.

Neural network architectures used in the modeling of the nervous systems can be divided into three categories, each of them having a different philosophy:

[5]Reshadat, V., and Feizi-Derakhshi, M.R., Neural Network-Based Methods in Information Retrieval, American Journal of Scientific Research, 2012, 58, 33–43.

[6]Bashiri, H., Neural Networks for Information Retrieval, http://www.powershow.com/view1/1af079-ZDc1Z/Neural_Networks_for_Information_Retrieval_powerpoint_ppt_presentation, 2005.

(1) *feedforward neural networks*, for which the transformation of the input vectors into the output vectors is determined by the refining of the system parameters;
(2) *feedback neural networks* (*recurrent neural networks*), where the input information defines the initial activity status of the feedback system and after passing through intermediate states, the asymptotic final state is identified as the result of the computation;
(3) *self-organizing maps* (that are introduced by Kohonen), within which the neighboring cells communicate with each other.

NNs work very well for IR from large text to multimedia databases; they have been widely used [5] in the area of IR and text mining, like text classification, text clustering, and collaborative filtering.

In recent years, with the fast growth of the World Wide Web and the Internet, these algorithms have also been used in Web-related applications such as Web searching, Web page clustering, Web mining, etc.

The capacity for tolerant and intuitive processing of the NNs offers new perspectives in IR [4, 5, 9, 10].

Using NNs in IR has [9] the following advantages:

- when the requested data (keywords) are not in the document collection, the NNs can be used to retrieve the information proximity around the required information;
- the information can be classified as the common patterns.

Model of the IRS with NNs "comes from the model based on statistical, linguistic and knowledge-based approach, which expresses document content and document relevance."[7]

Various neural networks such as Kohonen's Self-Organizing Map, Hopfield net, etc., have been applied [5] to IR models.

In the paper [11], we emphasized the possibility of using neural networks in IR and highlighted the advantages of applying two neural networks models for solving the problem of simplifying the complex structure of an IR system by substitution of the relations between its subsystems by NNs.

In this work, we reduced the text documents based on the Discrete Cosine Transformation (DCT), by which the set of keywords reduce to the much smaller feature set.

IR is a branch of Computing Science that aims at storing and allowing fast access to a large amount of information.

Technically, IR studies the acquisition, organization, storage, retrieval, and distribution of information.

IR is different from data retrieval in databases using SQL queries because the data in databases are highly structured and stored in relational tables, while information in text is unstructured.

There is no structured query language like SQL for text retrieval.

---

[7] Mokriš, I., and Skovajsová, L., Neural Network Model of System for Information Retrieval from Text Documents in Slovak Language, Acta Electrotechnica et Informatica, 2005, 3(5), 1–6.

1.2 Mathematical Background

1.2.1 Discrete Cosine Transformation

For large document collections, the high dimension of the vector space matrix F causes problems in text document set representation and high computing complexity in IR.

The most often methods of the text document space dimension reduction that have been applied in IR are the Singular Value Decomposition (SVD) and Principal Component Analysis (PCA).

The manner by which we shall reduce the text documents in this work is discrete cosine transformation, by which the set of keywords reduce to the much smaller feature set. The achieved model represents the latent semantic model.

The Discrete Cosine Transformation (DCT) [12] is an orthogonal transformation as the PCA, and the elements of the transformation matrix are computed using the following formula:

$$t_{mi} = \sqrt{\frac{2-\delta_{m-1}}{n}} \cos\left(\frac{\pi}{n}(i - \frac{1}{2}(m-1))\right), \ (\forall)\ i, m = \overline{1, n}, \tag{1.1}$$

n being the size of the transformation and

$$\delta_m = \begin{cases} 1 \text{ if } m = 1, \\ 0 \text{ otherwise.} \end{cases} \tag{1.2}$$

The DCT requires the transformation of the vectors $X_p,\ p = \overline{1, N}$ (N represents the number of vectors that must be transformed), of dimension n, to the vectors $Y_p,\ (\forall)\ p = \overline{1, N}$, according to the formula:

$$Y_p = T \cdot X_p,\ (\forall)\ p = \overline{1, N}, \tag{1.3}$$

$T = \{t_{mi}\}_{i,m=\overline{1,n}}$ meaning the transformation matrix.

Among all the components of the vectors $Y_p,\ p = 1, N$, we have to choose only m components, those from the positions for which one obtains a mean square that is in the first m mean squares, after they have been ordered descending, while the others $n - m$ components will be canceled.

If the vector by the index $p,\ p = \overline{1, N}$, achieved through the formula (1.3) is:

$$Y_p = \begin{pmatrix} y_{p1} \\ \vdots \\ y_{pn} \end{pmatrix},$$

then the mean square of the transformed vectors is

$$E(Y_p^2) = \begin{pmatrix} E(y_{p1}^2) \\ \vdots \\ E(y_{pn}^2) \end{pmatrix}, \tag{1.4}$$

where

$$E(y_{pj}^2) = \overline{y_{pj}^2} = \frac{1}{N} \sum_{p=1}^{N} y_{pj}^2, \ (\forall)\, j = \overline{1, n}.$$

The DCT application involves to determine the vectors $\hat{Y}_p,\ p = \overline{1, N}$ that contain those m components of the vectors $Y_p,\ p = \overline{1, N}$ that don't cancel.

1.2.2 *Algorithm for Image Compression Using Discrete Cosine Transformation*

Digital image processing involves [13–15] a succession of hardware and software processing steps and implementation of the theoretical methods.

The first stage of this process is the *image acquisition*, which requires an image sensor. It can be, for example, a video camera (like the pinhole camera model, one of the simplest camera models), which contributes to the achieving of a two-dimensional image.

The analog signal (being continuous in time and values) resulted at the output of the video camera has to be converted into a digital signal (in order to process it by means of the computer); this transformation involves the following three stages [14]:

Step 1 (*Spatial sampling*). This step has the aim to make the spatial sampling of the continuous light distribution. The spatial sampling of an image means the conversion of the continuous signal to its discrete representation and depends on the geometry of the sensor elements corresponding to the acquisition device.

Step 2 (*Temporal sampling*). During this step, the resulting discrete function has to be sampled in the time domain to create a single image. The temporal sampling is performed by measuring at regular intervals the amount of light, which is incident on each individual sensor element.

Step 3 (*Quantization of pixel values*). The purpose of present step is to quantize the resulting values of the image to a finite set of numeric values in order to store and process the image values on the computer.

Definition 1.1 ([14]). A *digital image* I is a two-dimensional function of natural coordinates $(u, v) \in \mathrm{N} \times \mathrm{N}$, which maps to a range of possible image (pixel) values P, such that $I(u, v) \in P$.

The pixel values are binary words of length k (which is called the *depth* of the image), so that a pixel can represent any of 2^k different values.

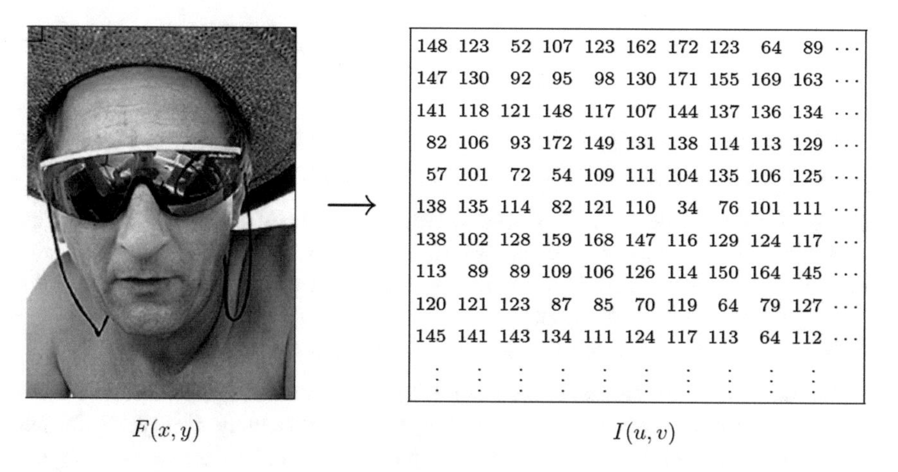

Fig. 1.3 Transformation of a continuous intensity function $F(x, y)$ to a discrete digital image $F(u, v)$ (see footnote 8)

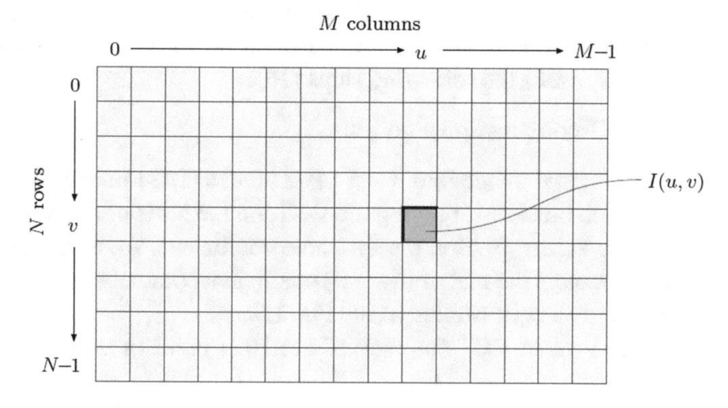

Fig. 1.4 Image coordinates (see footnote 8)

For example, the pixels of the grayscale images:

- are represented using $k = 8$ bits (1 byte) per pixel;
- have the intensity values in the set $\{0, 1, \ldots, 255\}$, where the value 0 represents the minimum brightness (black) and 255 the maximum brightness (white).

The result of the three steps *Step 1–Step 3* is highlighted in a "description of the image in the form of a two-dimensional, ordered matrix of integers"[8] (Fig. 1.3).

Figure 1.4 shows a coordinate system in the image processing, which is flipped in the vertical direction such that the origin ($u = 0,\ v = 0$) lies in the upper left corner.

[8]Burgerr, W., and Burge, M.J., Principles of Digital Image Processing. Fundamental Techniques, Springer-Verlag London, 2009.

The coordinates u, v represent the columns and the rows of the image, respectively. In the case of an image with dimensions $M \times N$, the maximum column number is $u_{\max} = M - 1$ and the maximum row number is $v_{\max} = N - 1$.

After obtaining the digital image it is necessary for its preprocessing in order to improve it; some examples of preprocessing techniques for images are:

1. *image enhancement*, which involves the transformation of the images to highlight: some hidden or obscure details, interest features, etc.;
2. *image compression*, made to reduce the amount of data required to represent a given amount of information;
3. *image restoration* corrects that errors that appear at the image capture.

Although there are different methods for image compression, the Discrete Cosine Transformation (DCT) achieves a good compromise between the ability of information compacting and the computational complexity. Another advantage of using the DCT in the image compression consists of the fact that it does not depend on the input data.

The DCT algorithm, which is used for the compression of 256×256, represented by the matrix of integers $X = (x_{ij})_{i,j=\overline{1,256}}$, where $x_{ij} \in \{0, 1, \ldots, 256\}$ means the original pixel values, needs the following steps [16]:

Algorithm 1.1 DCT compression algorithm.

Step 1 Split the original image into 8×8 pixel blocks (1024 image blocks).

Step 2 Process each block by applying the DCT, on the basis of the relation (1.3).

Step 3 Retain in a zigzag fashion the first nine coefficients for each transformed block and cancel the rest of $(64 - 9)$ coefficients (namely make them to be equal to 0) as it is illustrated in the Fig. 1.5.

Step 4 Apply the inverse DCT for each of the 1024 blocks resulted at previous step.

Step 5 Achieve the compressed image represented by the matrix $\hat{X} = (\hat{x}_{ij})_{i,j=\overline{1,256}}$, where $\hat{x}_{ij}$ denote the encoded pixel values and convert them into integer values.

Step 6 Evaluate the performances of the DCT compression algorithm in terms of the *Peak Signal-to-Noise Ratio* (PSNR), given by [16, 17]:

$$PSNR = 20 \log_{10} \frac{255}{\sqrt{MSE}}, \tag{1.5}$$

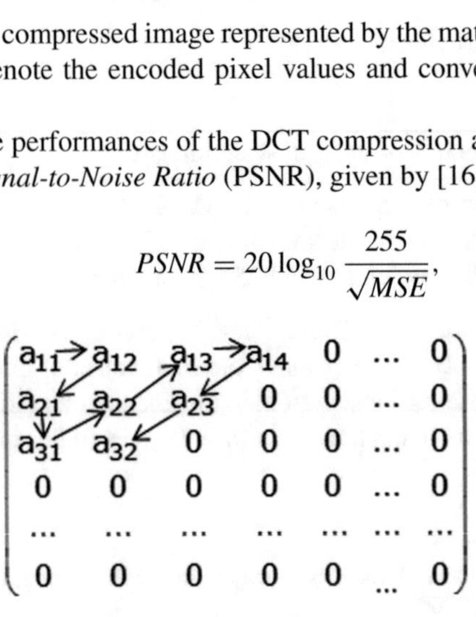

$$\begin{pmatrix} a_{11} & a_{12} & a_{13} & a_{14} & 0 & \ldots & 0 \\ a_{21} & a_{22} & a_{23} & 0 & 0 & \ldots & 0 \\ a_{31} & a_{32} & 0 & 0 & 0 & \ldots & 0 \\ 0 & 0 & 0 & 0 & 0 & \ldots & 0 \\ \ldots & \ldots & \ldots & \ldots & \ldots & \ldots & \ldots \\ 0 & 0 & 0 & 0 & 0 & \ldots & 0 \end{pmatrix}$$

Fig. 1.5 Zigzag fashion to retain the first nine coefficients

where the Mean Squared Error (MSE) is defined as follows [16, 17]:

$$MSE = \frac{1}{N \times N} \sum_{i=1}^{N} \sum_{j=1}^{N} (x_{ij} - \hat{x}_{ij})^2, \tag{1.6}$$

$N \times N$ being the total number of the pixels in the image (for our case $N = 256$).

We have experimented the compression algorithm using the DCT for the Lena.bmp image [16], which has 256×256 pixels and 256 levels of gray; it is represented in Fig. 1.6.

The Table 1.1 and Fig. 1.7 yields the experimental results achieved by implementing in MATLAB the DCT compression algorithm.

Fig. 1.6 Image Lena.bmp (see http://www.cosy.sbg.ac.at/~pmeerw/Watermarking/lena.html)

Table 1.1 Experimental results achieved by implementing the DCT compression algorithm

Number of the retained coefficients	Performances of the DCT compression algorithm	
	PSNR	MSE
$m_1 = 15$	27.6054	10.6232
$m_1 = 9$	25.3436	13.7836
$m_1 = 2$	21.2242	22.1477

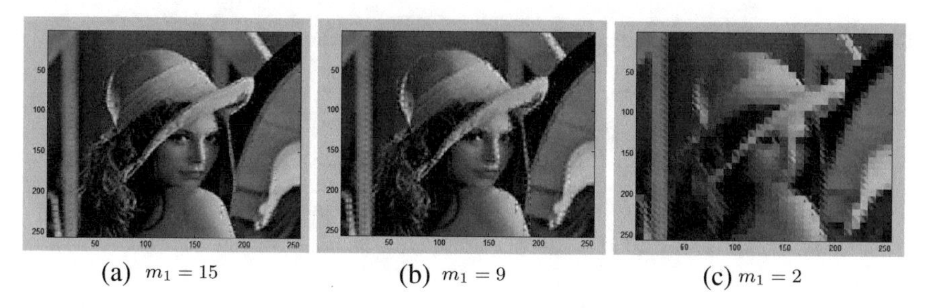

(a) $m_1 = 15$ (b) $m_1 = 9$ (c) $m_1 = 2$

Fig. 1.7 Visual evaluation of the performances corresponding to the DCT compression algorithm

1.2.3 Multilayer Nonlinear Perceptron

The Nonlinear Perceptron (NP) is the simplest, the most used and also, the oldest neural network. The name of "perceptron" derives from *perception*.

The classical structure of a NP (it is [18] a feedforward neural network, which contains three layers of neurons) is represented in Fig. 1.8.

The NP which has more than one hidden layer is called the Multilayer Nonlinear Perceptron (MNP).

The first layer of the MLP contains some virtual neurons, which does not perform a signal processing but only a multiplexing, the processing taking place only in the hidden and the output layer.

The Equations of the Neurons from the Hidden Layer

The Fig. 1.9 shows the connections of a neuron from the hidden layer.

- $X_p = (x_{p1}, \ldots, x_{pn})$ signifies a vector, which is applied to the NP input;
- $\{W^h_{ji}\}_{j=\overline{1,L},\ i=\overline{1,n}}$ represents the set of the weights corresponding to the hidden layer;
- L is the number of the neurons belonging to the hidden layer;
- n represents the number of the neurons from output layer.

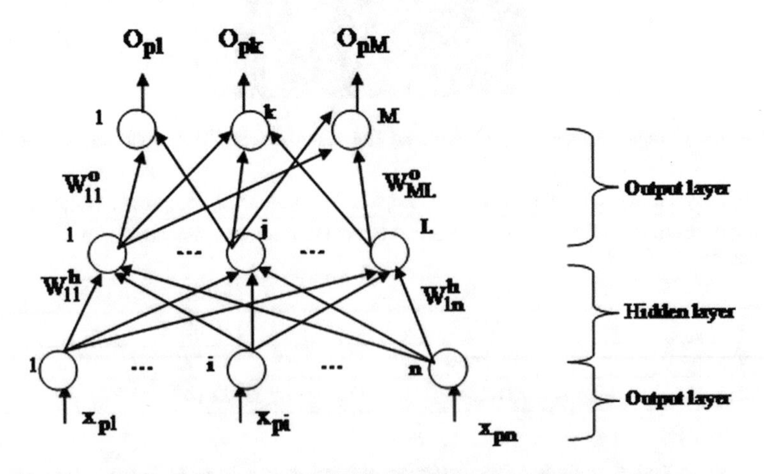

Fig. 1.8 The classical structure of a NP with three layers of neurons

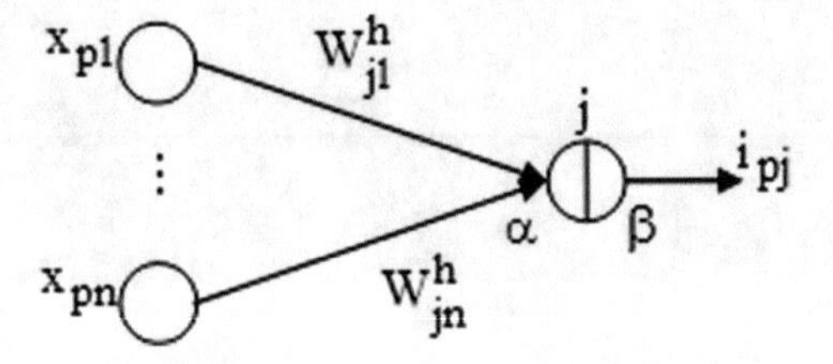

Fig. 1.9 The connections of the neuron j ($j = \overline{1, L}$) from the hidden layer

The upper index "h" which occurs in the notation of the weights afferent to this layer comes from "hidden".

There are two processing stages:

(α) linear stage;
(β) nonlinear stage.

(α) Linear Stage

During this stage, we shall define the activation of the neuron j from the hidden layer, when the vector X_p is applied at the network input.

$$(\alpha) \quad net_{pj}^h = \sum_{i=1}^{n} W_{ji}^h \cdot x_{pi}, \ (\forall)\, j = \overline{1,\ L}, \tag{1.7}$$

where net_{pj}^h means the activation of the neuron j.

If

$$W_j^h = (W_{j1}^h, \ldots, W_{jn}^h)^t,$$

then

$$net_{pj}^h = < W_j^h, X_p > = (W_j^h)^t \cdot X_p, \ (\forall)\, j = \overline{1,\ L}, \tag{1.8}$$

where $< , >$ means the scalar product and t signifies the transpose operation.

(β) Nonlinear Stage

This stage involves the calculation of hidden neuron j, according to the formula:

$$(\beta) \quad i_{pj}^h = f_j^h(net_{pj}^h), \ (\forall)\, j = \overline{1,\ L}, \tag{1.9}$$

where f_j^h is a nonlinear function by the type:

$$f : \Re \rightarrow \Re, \ f(x) = \frac{1}{1 + e^{-\lambda x}} \ (\lambda > 0 \textit{ is a constant});$$

for example f can be a nonlinear sigmoid (see the Fig. 1.10) or $f(x) = \text{th}\ x$ (the nonlinear hyperbolic tangent function), see the Fig. 1.11.

The Equations of the Neurons from the Output Layer

The Fig. 1.12 represents the connections of a neuron from the output layer.

- M constitutes the number of the neurons belonging to the output layer;
- $\{W_{kj}^o\}_{k=\overline{1,M},\ j=\overline{1,L}}$ represents the set of the weights corresponding to the output layer (the index "o" comes from "output").

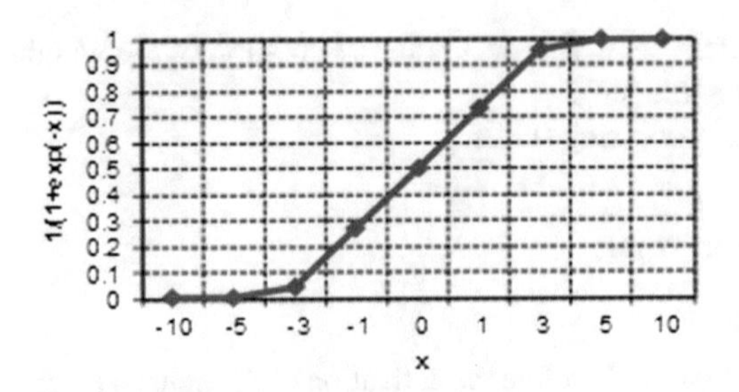

Fig. 1.10 The nonlinear sigmoid function

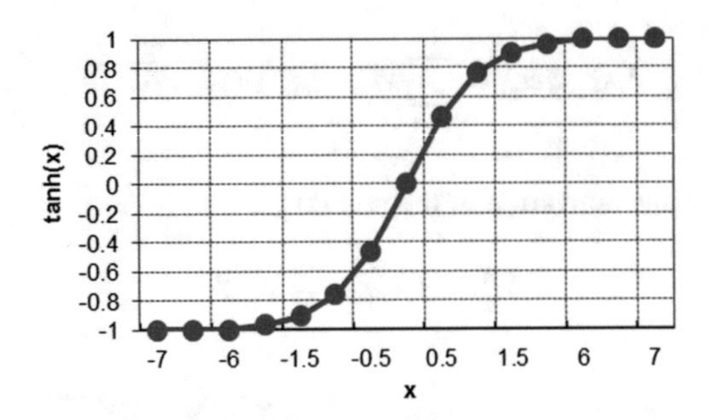

Fig. 1.11 The nonlinear hyperbolic tangent function

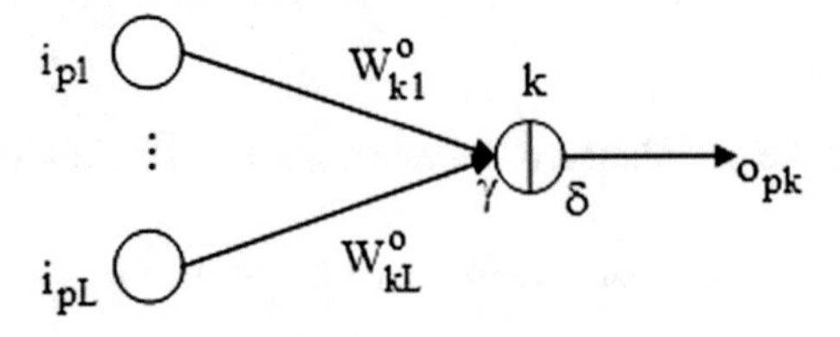

Fig. 1.12 The connections of the neuron k ($k = \overline{1, M}$) from the output layer

There are two processing stages:

(γ) linear stage;
(δ) nonlinear stage.

(γ) Linear Stage

By applying the vector X_p at the input network, the activation of the neuron k from the output layer is given by the formula:

$$(\gamma) \quad net^o_{pk} = \sum_{j=1}^{L} W^o_{kj} \cdot i_{pj}, \; (\forall) \; k = \overline{1, \; M}, \tag{1.10}$$

where net^o_{pk} is the activation of the neuron k.

(δ) Nonlinear Stage

The output of neuron k will be determined like a function of the activation net_{pk}^{o} and it is expressed by the relation:

$$(\delta) \quad o_{pk} = f_k^o(net_{pk}^o), \; (\forall)\, k = \overline{1,\, M}, \tag{1.11}$$

where $f_k^o(net_{pk}^o)$ can be a sigmoid or a hyperbolic tangent function.

The training algorithm of NP is supervised, by the backpropagation type, in order to minimize the error on the training lot,

$$E = \frac{1}{K}\sum_{p=1}^{K} E_p, \tag{1.12}$$

K being the number of the vectors from the training lot and

$$E_p = \frac{1}{2}\sum_{k=1}^{M} (y_{pk} - o_{pk})^2, \tag{1.13}$$

constitutes the error determined by the training vector, having the index p, where:

- $Y_p = (y_{p1}, \ldots, y_{pM})$ represents the ideal vector;
- $O_p = (o_{p1}, \ldots, o_{pM})$ signifies the output real vector.

1.2.4 Fuzzy Neural Perceptron

We shall build two variants of the multilayer perceptron, that will be used for an IR model and are denoted with FNP (Fuzzy Nonlinear Perceptron) followed by a digit [19], which means the index version of the fuzzy perceptron:

(1) **FNP1**

In case when we used the first fuzzy variant FNP1, the input vectors will have dimension $5n$, resulting from the vectors of size n that have provided to the input of the classical perceptron, in which each component was represented by five values that constitute the membership of that component to the five linguistic terms: *unimportant*, *rather unimportant*, *moderately important*, *rather important*, and *very important*; they can be represented as fuzzy numbers (see Fig. 1.13). The outputs of the FNP1 will be non-fuzzy (crisp).

The membership functions of the fuzzy terms *unimportant*, *rather unimportant*, *moderately important*, *rather important*, and respectively *very important* are defined [20] in the following relations:

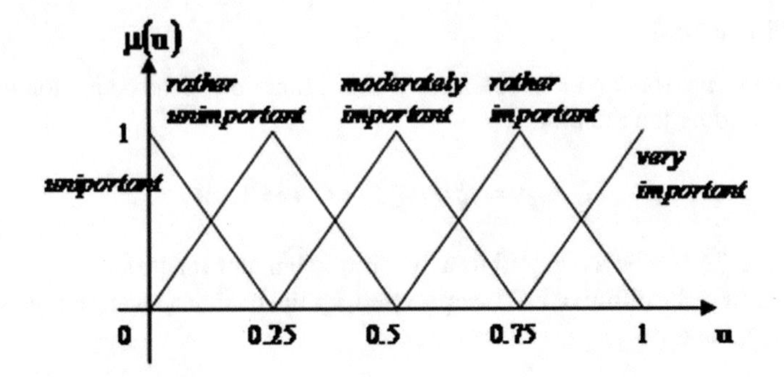

Fig. 1.13 The membership functions of *importance*

$$\mu_{unimportant} = \begin{cases} 1 - \frac{x}{0.25}, \text{ if } 0 \leq x \leq 0.25, \\ 0, \text{ otherwise.} \end{cases} \quad (1.14)$$

$$\mu_{rather\ unimportant} = \begin{cases} \frac{x}{0.25} \text{ if } 0 \leq x \leq 0.25, \\ 1 - \frac{x-0.25}{0.25} \text{ if } 0.25 < x \leq 0.5, \end{cases} \quad (1.15)$$

$$\mu_{moderately\ important} = \begin{cases} \frac{x-0.25}{0.25}, \text{ if } 0.25 \leq x \leq 0.5, \\ 1 - \frac{x-0.5}{0.25}, \text{ if } 0.5 < x \leq 0.75, \end{cases} \quad (1.16)$$

$$\mu_{rather\ important} = \begin{cases} \frac{x-0.5}{0.25}, \text{ if } 0.5 \leq x \leq 0.75, \\ 1 - \frac{x-0.75}{0.25}, \text{ if } 0.75 < x \leq 1, \end{cases} \quad (1.17)$$

$$\mu_{very\ important} = \begin{cases} \frac{x-0.75}{0.25}, \text{ if } 0.75 \leq x \leq 1, \\ 0, \text{ otherwise.} \end{cases} \quad (1.18)$$

(2) **FNP2**

An other version of the fuzzy perceptron has fuzzy its ideal outputs; for each vector which is applied to the network input, the ideal output vector is of the form:

$$(\mu(y_1), \ldots, \mu(y_M)),$$

each component being represented [21] by the nonlinear function:

$$\mu(y_i) = \frac{1}{1 + \left(\frac{z_{ik}}{F_d}\right)^{F_e}}, \ (\forall)\ k = \overline{1, K},\ i = \overline{1, M}, \tag{1.19}$$

where:

- K is the number of the vectors from the training lot and respectively from the test lot;
- M means the number of classes;
- z_{ik} is the weighted distance between each input vector

  $$X_k = (x_{k1}, x_{k2}, \ldots, x_{kn}), k = \overline{1, K}$$

  and the mean vector of each class

  $$Med_i = (med_{i1}, med_{i2}, \ldots, med_{in}), i = \overline{1, M},$$

  namely:

  $$z_{ik} = \sum_{h=1}^{n} (x_{kh} - med_{ih}),\ (\forall)\ k = \overline{1, K},\ i = \overline{1, M}, \tag{1.20}$$

  n being the dimension of the vectors;
- F_d and F_e are some parameters that control the amount of fuzziness in the class membership.

1.3 A New Approach of a Possibility Function Based Neural Network

Probabilistic neural network [22] is a kind of involved from the radial basis function networks, having the Bayesian minimum risk criteria as its theoretical basis. In pattern classification, its advantage consists of substituting the nonlinear learning algorithm with linear learning algorithm.

In recent years, probabilistic neural network are also used in the field of IR [23], face recognition [22], for its structure, good approximation, fast training speed and good real-time performance.

This section presents a new type of fuzzy neural network, entitled Possibility Function-based Neural Network (PFBNN). Its advantages consist of that it not only can perform as a standard neural network, but can also accept a group of possibility functions as input.

The PFBNN discussed [24] in this section has novel structures, consisting of two stages:

1. the first stage of the network is a fuzzy based and it has two parts: a Parameter Computing Network (PCN), followed by a Converting Layer (CL);
2. the second stage of the network is a standard backpropagation based neural network (BPNN).

The PCN in a possibility function-based network can also be used to predict functions. The CL is used to convert the possibility function to a value. This layer is necessary for data classification. The network can still function as a classifier using only the PCN and the CL or only the CL. Using only the PCN one can perform a transformation from one group of possibility functions to another.

1.4 Architecture of the PFBNN

From the Fig. 1.14, which shows the block diagram of the proposed fuzzy neural network we can notice that there are two stages:

(1) the first stage of the network is a fuzzy based and it has two parts: a Parameter Computing Network (PCN), followed by a Converting Layer (CL);
(2) the second stage of the network is a standard backpropagation based neural network (BPNN).

This PBFNN can be segmented and still perform useful functions. The network can still function as a classifier using only the PCN and the CL or only the CL. Using only the PCN one can perform a transformation from one group of possibility functions to another.

There are three types of weight variables used between connecting nodes of the three networks in Fig. 1.14:

a. the first type called a λ-weight is used for connection weights between nodes in the PCN;
b. the second type called a r-weight is used for connection weights between the output of the PCN and the CL;

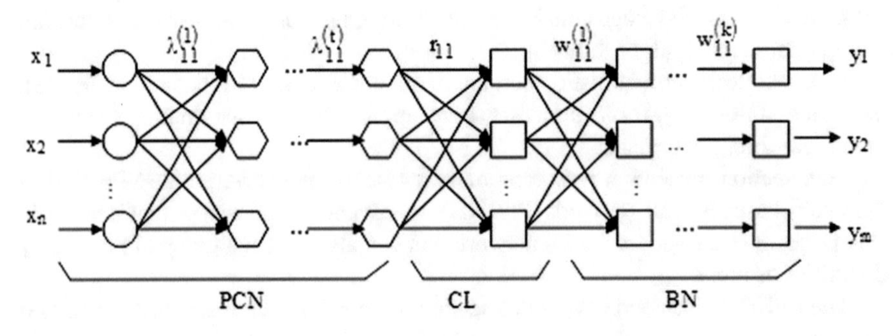

Fig. 1.14 The framework of a possibility-based fuzzy neural network

c. the third type called a w-weight is used for connection weights between the neurons in a standard BPNN.

As the w-weights are adjusted according to standard backpropagation algorithms, we shall discuss only the setting and adjustment of the λ- and r-weights.

The PCN accepts as input, a vector representing a group of possibility functions and generates a group of possibility functions as output.

In the PCNN, the weights associated of the neurons corresponding to the λ layers are

$$\{\lambda_{ij}^{(k)}\}_{i=\overline{1,L_{k-1}},\ j=\overline{1,L_k}}$$

where:

- k ($k = \overline{1,t}$) represents the order of the PCN layer,
- i is the index of the neuron from the $(k-1)$ layer,
- j means the index of the neuron from the k layer,
- L_k is the number of the neurons of the k-th layer, $L_0 = n$,

with

$$\lambda_{ij}^{(k)} : [0, 1] \rightarrow [-1, 1]$$

and they are always positive or always negative.

Each $\lambda_{ij}^{(k)}$ is represented as a binary tuple $(\rho_{ij}^{(k)},\ \omega_{ij}^{(k)})$, where:

- $\rho_{ij}^{(k)}$ is a transformation function from [0, 1] to [0, 1];
- $\omega_{ij}^{(k)}$ is a constant real number in $[-1, 1]$.

One can use a fuzzy normal distribution function

$$f(x) = \mathrm{e}^{-\frac{(x-\mu)^2}{2\sigma^2}} \tag{1.21}$$

for each element x of the crisp input vector X to obtain the fuzzified input data of the PCN.

We shall compute the outputs of the neurons from the k-th layer of PCN using the relation:

$$y_j^{(k)}(u) = \sum_{i=1}^{L_{k-1}} \omega_{ij}^{(k)} y_i^{(k-1)} \left(\rho_{ij}^{(k)}(u)^{-1}\right),\ u \in [0, 1] \tag{1.22}$$

or shortly

$$y_j^{(k)} = \sum_{i=1}^{L_{k-1}} \omega_{ij}^{(k)} \left(y_i^{(k-1)} \circ \rho_{ij}^{(k)^{-1}}\right), \tag{1.23}$$

where:

- $Y^{(k)} = (y_1^{(k)}, \ldots, y_{L_k}^{(k)})$ is the output vector of the k-th layer of PCN,
- $Y^{(k-1)} = (y_1^{(k-1)}, \ldots, y_{L_k-1}^{(k)})$ constitutes the input vector of the k-th layer (namely the output vector of the $(k-1)$-th layer),
- "$\circ$" means the composite of two functions,
- ${\rho_{ij}^{(k)}}^{-1}$ is the inverse function of $\rho_{ij}^{(k)}$.

The CL accepts as input a possibility function (that represents a possibility vector) generated by the PCN and transforming it into a real number vector.

Each weight of this layer is a function

$$r_{ij} : [0, 1] \to [-1, 1],\ i = \overline{1, L_t},\ j = \overline{1, M},$$

where L_t is the number of the neurons from the layer t of PCN and M is the number of the output neurons of CL.

Similar to λ in the PCN, r_{ij} is always positive or always negative. The r-weights of the CL can be also represented as a binary tuple $r_{ij} = (\gamma_{ij}, \tau_{ij}), i = \overline{1, L_t}, j = \overline{1, M}$ where

$$\gamma_{ij} : [0, 1] \to [0, 1]$$

is a possibility function, which is different from ρ (the transformation function in the PCN) and τ_{ij} is a constant real number in $[-1, 1]$.

The output $Z = (z_1, \ldots, z_M)$ of the CL is a vector having real numbers as components, see the following formula:

$$z_j = \sum_{i=1}^{L_t} \tau_{ij} \left(\max_{u \in [0,1]} \left(\min(y_i^{(t)}(u), \tau_{ij}(u)) \right) \right),\ j = \overline{1, M}, \tag{1.24}$$

$Y^{(t)} = (y_1^{(t)}, \ldots, y_{L_t}^{(t)})$ being the fuzzy input vector of CL, which constitutes the output vector of PCN.

We shall use $y_i^{(t)}(u) \cdot \tau_{ij}(u)$ instead of $\min(y_i^{(t)}(u), \tau_{ij}(u))$ in order to compute easier the outputs of the CL using (1.24).

1.5 Training Algorithm of the PBFNN

We shall build the training algorithm of the PBFNN in the hypothesis that the PCN has three layers (namely $t = 3$):

1. the input layer which contains a number of $L_0 = n$ neurons
2. a hidden layer having L_1 neurons
3. an output layer with L_2 neurons.

Step 1 Initialize the weights of the PCN and CL in the following way:

(a) choose a linear function as the initial weight function for each ρ:

$$\rho_{ij}^{(k)}(u_i) = v_j,\ i = \overline{1, L_t},\ j = \overline{1, M} \tag{1.25}$$

and design a genetic algorithm to search for optimal ω's;

(b) let each weight function γ as a possibility function:

$$\gamma_{ij}(u) = \mathrm{e}^{-\frac{(u-u_0)^2}{2\sigma^2}},\ u_0 \in [0, 1],\ i = \overline{1, L_t},\ j = \overline{1, M} \tag{1.26}$$

assigning usually $\sigma = 1$ and design a genetic algorithm to search for optimal τ's.

Let $Y^{(0)} = (y_1^{(0)}, \ldots, y_{L_0}^{(0)})$ be the input fuzzy vector of the PCN corresponding to the training vector by the index p.

Step 2 Compute the fuzzy output vector $Y^{(1)} = (y_1^{(1)}, \ldots, y_{L_1}^{(1)})$ of the hidden layer of PCN using the relation:

$$y_j^{(1)} = \sum_{i=1}^{L_0} \omega_{ij}^{(1)} \left(y_i^{(0)} \circ \rho_{ij}^{(1)^{-1}} \right),\ j = \overline{1, L_1}. \tag{1.27}$$

Step 3 Compute the fuzzy output vector $Y^{(2)} = (y_1^{(2)}, \ldots, y_{L_2}^{(1)})$ of the output layer of PCN using the relation:

$$y_k^{(2)} = \sum_{j=1}^{L_1} \omega_{jk}^{(2)} \left(y_j^{(1)} \circ \rho_{jk}^{(2)^{-1}} \right),\ k = \overline{1, L_2}. \tag{1.28}$$

Step 4 Apply to the input of the CL the fuzzy vector $Y^{(2)} = (y_1^{(2)}, \ldots, y_{L_2}^{(1)})$, which one obtains at the output of the PCN.

Step 5 Determine the output vector $Z = (z_1, \ldots, z_M)$ of the CL, having each component a real number:

$$z_j = \sum_{i=1}^{L_2} \tau_{ij} \left(\max_{u \in [0,1]} \left(y_i^{(2)}(u), \tau_{ij}(u) \right) \right),\ j = \overline{1, M}, \tag{1.29}$$

where M is the number of the output neurons of the CL.

Step 6 Adjust the weights of the output layer of the PCN:

$$\begin{cases} \rho_{jk}^{(2)}(u_j) \leftarrow \rho_{jk}^{(2)}(u_j) + \mu_\rho \cdot \dfrac{\partial E}{\partial \rho_{jk}^{(2)}(u_j)} \\ \omega_{jk}^{(2)}(u_j) \leftarrow \omega_{jk}^{(2)}(u_j) + \mu_\omega \cdot \dfrac{\partial E}{\partial \omega_{jk}^{(2)}(u_j)}, \end{cases} \tag{1.30}$$

$j = \overline{1, L_1}$, $k = \overline{1, L_2}$, μ_ρ, μ_ω being two constants with the meaning of learning rates, and

$$E = \frac{1}{|S_T|} \sum_{p=1}^{|S_T|} E_p \tag{1.31}$$

defines the performance of the system, where:

- $|S_T|$ represents the number of the vectors from the training lot,
- E_p is the output error of the PCN for the p-th training sample, defined by:

$$E_p = \sum_{k=1}^{L_2} \left(\int_0^1 E_p(v_k) \mathrm{d}v_k \right) \tag{1.32}$$

and

$$E_p(v_k) = \frac{1}{2} \left(T_k(v_k) - y_k^{(2)}(v_k) \right)^2, \tag{1.33}$$

$T_k = (T_1, \ldots, T_{L_2})$ being the ideal output vector (the target vector) of the input vector by the index p applied to the PCN.
We shall have:

$$\frac{\partial E}{\partial \rho_{jk}^{(2)}(u_j)} = \sum_{k=1}^{L_2} \left(\int_0^1 \frac{\partial E_p(v_k)}{\partial \rho_{jk}^{(2)}(u_j)} dv_k \right) = \int_0^1 \frac{\partial E_p(v_k)}{\partial \rho_{jk}^{(2)}(u_j)} \mathrm{d}v_k =$$

$$= - \int_0^1 \left(T_k(v_k) - y_k^{(2)}(v_k) \right) \cdot \frac{\partial y_k^{(2)}(v_k)}{\partial \rho_{jk}^{(2)}(u_j)} \mathrm{d}v_k \tag{1.34}$$

and

$$\frac{\partial y_k^{(2)}(v_k)}{\partial \rho_{jk}^{(2)}(u_j)} = \omega_{jk}^{(2)} \cdot \frac{\partial y_j^{(1)}}{\partial {\rho_{jk}^{(2)}}^{-1}(v_k)} \cdot \frac{\partial {\rho_{jk}^{(2)}}^{-1}(v_k)}{\partial \rho_{jk}^{(2)}(u_j)}. \tag{1.35}$$

Substituting (1.34) and (1.35) into (1.30) it will results:

$$\rho_{jk}^{(2)}(u_j) \leftarrow \rho_{jk}^{(2)}(u_j) - \mu_\rho \omega_{jk}^{(2)} \cdot$$

$$\int_0^1 \left(T_k(v_k) - y_k^{(2)}(v_k) \right) \cdot \frac{\partial y_j^{(1)}}{\partial {\rho_{jk}^{(2)}}^{-1}(v_k)} \cdot \frac{\partial {\rho_{jk}^{(2)}}^{-1}(v_k)}{\partial \rho_{jk}^{(2)}(u_j)} \mathrm{d}v_k, \tag{1.36}$$

where $i = \overline{1, L_1}$, $k = \overline{1, L_2}$.

Similarly,

$$\frac{\partial E}{\partial \omega_{jk}^{(2)}} = \sum_{k=1}^{L_2} \left(\int_0^1 \frac{\partial E_p(v_k)}{\partial \omega_{jk}^{(2)}(u_j)}) dv_k \right) = \int_0^1 \frac{\partial E_p(v_k)}{\partial \omega_{jk}^{(2)}} \mathrm{d}v_k \tag{1.37}$$

namely

$$\frac{\partial E}{\partial \omega_{jk}^{(2)}} = - \int_0^1 \left(T_k(v_k) - y_k^{(2)}(v_k) \right) \cdot \frac{\partial y_k^{(2)}(v_k)}{\partial \omega_{jk}^{(2)}} \mathrm{d}v_k, \tag{1.38}$$

where

$$\frac{\partial y_k^{(2)}(v_k)}{\partial \omega_{jk}^{(2)}} = y_j^{(1)} \left(\rho_{jk}^{(2)^{-1}}(v_k) \right). \tag{1.39}$$

Substituting (1.38) and (1.39) into (1.30) we obtain:

$$\omega_{jk}^{(2)}(u_j) \leftarrow \omega_{jk}^{(2)}(u_j) - \mu_\omega \int_0^1 \left(T_k(v_k) - y_k^{(2)}(v_k) \right) \cdot y_j^{(1)} \left(\rho_{jk}^{(2)^{-1}}(v_k) \right) \mathrm{d}v_k \tag{1.40}$$

where $i = \overline{1, L_1}$, $k = \overline{1, L_2}$.

Step 7 Adjust the weights of the hidden layer of the PCN:

$$\begin{cases} \rho_{ij}^{(1)}(u_i) \leftarrow \rho_{ij}^{(1)}(u_i) + \mu_\rho \cdot \frac{\partial E}{\partial \rho_{ij}^{(1)}(u_i)} \\ \omega_{ij}^{(1)}(u_i) \leftarrow \omega_{ij}^{(1)}(u_i) + \mu_\omega \cdot \frac{\partial E}{\partial \omega_{ij}^{(1)}(u_i)}, \end{cases} \tag{1.41}$$

$i = \overline{1, L_0}$, $j = \overline{1, L_1}$, where:

$$\frac{\partial E}{\partial \rho_{ij}^{(1)}(u_i)} = \sum_{k=1}^{L_2} \left(\int_0^1 \frac{\partial E_p(v_k)}{\partial \rho_{ij}^{(1)}(u_i)}) dv_k \right), \tag{1.42}$$

namely:

$$\frac{\partial E}{\partial \rho_{ij}^{(1)}(u_i)} = - \sum_{k=1}^{L_2} \left(\int_0^1 (T_k(v_k) - y_k^{(2)}(v_k)) \cdot \frac{\partial y_k^{(2)}(v_k)}{\partial \rho_{ij}^{(1)}(u_i)} dv_k \right) \tag{1.43}$$

where

$$y_k^{(2)}(v_k) = \sum_{j=1}^{L_1} \omega_{jk}^{(2)} \cdot y_j^{(1)} \left(\rho_{jk}^{(2)^{-1}}(v_k) \right), \tag{1.44}$$

namely

$$y_k^{(2)}(v_k) = \sum_{j=1}^{L_1} \omega_{jk}^{(2)} \cdot \sum_{i=1}^{L_0} \omega_{ij}^{(1)} \cdot y_i^{(0)} \left(\rho_{ij}^{(1)^{-1}}(\rho_{jk}^{(2)^{-1}}(v_k))\right) \tag{1.45}$$

and

$$\frac{\partial y_k^{(2)}(v_k)}{\partial \rho_{ij}^{(1)}(u_i)} = \omega_{jk}^{(2)} \cdot \omega_{ij}^{(1)} \cdot \frac{\partial y_i^{(0)}}{\partial \rho_{ij}^{(1)^{-1}}(\rho_{jk}^{(2)^{-1}}(v_k))} \cdot \frac{\partial \rho_{ij}^{(1)^{-1}}(\rho_{jk}^{(2)^{-1}}(v_k))}{\rho_{ij}^{(1)}(u_i)}, \tag{1.46}$$

where

$$\frac{\partial \rho_{ij}^{(1)^{-1}}(\rho_{jk}^{(2)^{-1}}(v_k))}{\rho_{ij}^{(1)}(u_i)} = \frac{\partial \rho_{ij}^{(1)^{-1}}}{\partial \rho_{jk}^{(2)^{-1}}(v_k)} \cdot \frac{\partial \rho_{jk}^{(2)^{-1}}(v_k)}{\partial \rho_{ij}^{(1)}(u_i)} \tag{1.47}$$

Substituting (1.43), (1.46), (1.47) into the first formula from the relation (1.41) we achieve:

$$\rho_{ij}^{(1)}(u_i) \leftarrow \rho_{ij}^{(1)}(u_i) - \mu_\rho \cdot \omega_{ij}^{(1)} \cdot$$

$$\sum_{k=1}^{L_2} \omega_{jk}^{(2)} \left(T_k(v_k) - y_k^{(2)}(v_k)\right) \cdot \frac{\partial y_i^{(0)}}{\partial \rho_{ij}^{(1)^{-1}}(\rho_{jk}^{(2)^{-1}}(v_k))} \cdot \frac{\partial \rho_{ij}^{(1)^{-1}}}{\partial \rho_{jk}^{(2)^{-1}}(v_k)} \cdot \frac{\partial \rho_{jk}^{(2)^{-1}}(v_k)}{\partial \rho_{ij}^{(1)}(u_i)} \mathrm{d}v_k, \tag{1.48}$$

$(\forall)\ i = \overline{1, L_0}, j = \overline{1, L_1}$.
Analogically,

$$\frac{\partial E}{\partial \omega_{ij}^{(1)}(u_i)} = \sum_{k=1}^{L_2} \left(\int_0^1 \frac{\partial E_p(v_k)}{\partial \omega_{ij}^{(1)}(u_i)})\mathrm{d}v_k \right) \tag{1.49}$$

namely

$$\frac{\partial E}{\partial \omega_{ij}^{(1)}(u_i)} = -\sum_{k=1}^{L_2} \left(\int_0^1 (T_k(v_k) - y_k^{(2)}(v_k)) \cdot \frac{\partial y_k^{(2)}(v_k)}{\partial \omega_{ij}^{(1)}(u_i)} \mathrm{d}v_k \right), \tag{1.50}$$

where

$$\frac{\partial y_k^{(2)}(v_k)}{\partial \omega_{ij}^{(1)}(u_i)} = \omega_{jk}^{(2)} \cdot y_i^{(0)} \left(\rho_{ij}^{(1)^{-1}}(\rho_{jk}^{(2)^{-1}}(v_k))\right). \tag{1.51}$$

Substituting (1.50), (1.51) into the second formula from the relation (1.41) we achieve:

$$\omega_{ij}^{(1)}(u_i) \leftarrow \omega_{ij}^{(1)}(u_i) - \mu_\omega \cdot$$

$$\sum_{k=1}^{L_2} \omega_{jk}^{(2)} \int_0^1 \left(T_k(v_k) - y_k^{(2)}(v_k)\right) \cdot y_i^{(0)} \left(\rho_{ij}^{(1)^{-1}}(\rho_{jk}^{(2)^{-1}}(v_k))\right) \mathrm{d}v_k, \tag{1.52}$$

$(\forall)\ i = \overline{1, L_2}, j = \overline{1, M}$.

Step 8 Adjust the weights of the CL:

$$\begin{cases} \gamma_{ij}(u) \leftarrow \gamma_{ij}(u) + \mu_\gamma \cdot \frac{\partial E}{\partial \gamma_{ij}(u)} \\ \tau_{ij} \leftarrow \tau_{ij} + \mu_\tau \cdot \frac{\partial E}{\partial \tau_{ij}}, \end{cases} \tag{1.53}$$

$u \in [0, 1]$, $(\forall)\ i = \overline{1, L_2}, j = \overline{1, M}$, μ_γ and μ_τ being two constants with the meaning of learning rates and E is the performance of the system (defined as in (1.31)), E_p being in this case the output error of the CL for the p-th training sample and it is defined by:

$$E_p = \frac{1}{2} \sum_{j=1}^{M} (U_j - z_j)^2, \tag{1.54}$$

$U = (U_1, \ldots, U_M)$ being the ideal output vector (the target vector) of the input vector by the index p applied to the CL.
We shall have:

$$\frac{\partial E}{\partial \gamma_{ij}(u)} = \frac{\partial E_p}{\partial \gamma_{ij}(u)} = -(U_j - z_j) \frac{\partial z_j}{\partial \gamma_{ij}(u)}. \tag{1.55}$$

Let $u_{\max}$ the point for which $y_i^{(2)}(u)\gamma_{ij}(u)$ has maximum value.
Hence,

$$\frac{\partial z_j}{\partial \gamma_{ij}(u)} = \sum_{k=1}^{M} \tau_{kj} \cdot \frac{\partial \left(y_k^{(2)}(u) \cdot \gamma_{kj}(u)\right)}{\partial \gamma_{ij}(u_{\max})}, \tag{1.56}$$

namely:

$$\frac{\partial z_j}{\partial \gamma_{ij}(u)} = \sum_{k=1}^{M} \tau_{kj} \left(\gamma_{kj}(u_{\max}) \cdot \frac{\partial y_k^{(2)}}{\partial \gamma_{ij}(u_{\max})} + y_k^{(2)}(u_{\max}) \cdot \frac{\partial \gamma_{kj}}{\partial \gamma_{ij}(u_{\max})} \right), \tag{1.57}$$

where

$$\frac{\partial y_k^{(2)}}{\partial \gamma_{ij}(u_{\max})} = 0 \tag{1.58}$$

and

$$\frac{\partial \gamma_{kj}}{\partial \gamma_{ij}(u_{\max})} = \begin{cases} 1, & \text{if } k = i \\ 0, & \text{otherwise.} \end{cases} \tag{1.59}$$

Introducing the relations (1.56)–(1.59) into the first formula from (1.53) one obtains:

$$\gamma_{ij}(u_{\max}) \leftarrow \gamma_{ij}(u_{\max}) - \mu_\gamma \cdot \tau_{ij} \cdot (U_j - z_j) \cdot y_i^{(2)}(u_{\max}), \tag{1.60}$$

$(\forall)\ i = \overline{1, L_2}, j = \overline{1, M}$.
Similarly, we shall have:

$$\frac{\partial E}{\partial \tau_{ij}} = \frac{\partial E_p}{\partial \tau_{ij}} = -(U_j - z_j)\frac{\partial z_j}{\partial \tau_{ij}}, \tag{1.61}$$

where

$$\frac{\partial z_j}{\partial \tau_{ij}} = \max_{u \in [0,1]} \left(y_i^{(2)}(u) \cdot \gamma_{ij}(u) \right) = y_i^{(2)}(u_{\max}) \cdot \gamma_{ij}(u_{\max}). \tag{1.62}$$

Substituting (1.61) and (1.62) into the second formula from (1.53) we shall achieve:

$$\tau_{ij} \leftarrow \tau_{ij} - \mu_\tau \cdot \tau_{ij} \cdot (U_j - z_j) \cdot y_i^{(2)}(u_{\max}) \cdot \gamma_{ij}(u_{\max}), \tag{1.63}$$

$(\forall)\ i = \overline{1, L_2}, j = \overline{1, M}$.

Step 9 Compute the PCN error because of the p-th training vector with (1.32).

Step 10 Compute the CL error because of the p-th training vector, using (1.54).

Step 11 If the training algorithm has not applied for all the training vectors, then go to the next vector.
Otherwise, test the stop condition. For example, we can stop the algorithm after a fixed training epoch numbers.

1.6 Neural Networks-Based IR

NNs are "powerful techniques for representing complex relationships between inputs and outputs. Based on the neural structure of the brain, NNs are complicated and they can be enormous for certain domains, containing a large number of nodes and synapses."[9]

---

[9] Xhemali, D., and Hinde, C.J., and Stone, R.G., Naïve Bayes vs. Decision Trees vs. Neural Networks in the Classification of Training Web Pages, International Journal of Computer Science Issues, 2009, 4(1), 16–23.

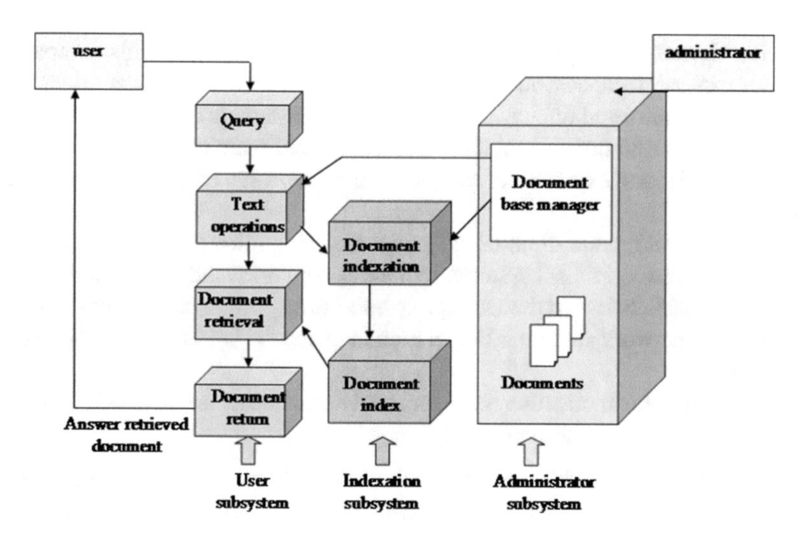

Fig. 1.15 A general architecture of an IR system

Problem of the text document retrieval can be solved also by NNs. Nowadays, there are different NNs used for text document retrieval [5, 9, 10, 25]. Generally, one can divide NNs for text document retrieval into three main categories: feedforward neural networks [26], Kohonen neural networks [27] and recurrent neural networks.[10]

Figure 1.15 shows a general architecture of an IR system.

An IR system has [26] three main subsystems as it is on the Fig. 1.15: first subsystem is the user query subsystem where user poses a query, second is the indexing subsystem where the documents and query are compared, and third is the administrator or documents subsystem where the documents are stored. The administrator subsystem guaranties the administration of the documents. Administrator determines the document base from the document set. Document base manager aims to:

- provide the system representation of the documents;
- determines a suitable model for document storage and creates the system representation of the documents.

Indexation subsystem has the task of creating:

- an index;
- a query representation that is comparable with the document index.

The aim of the user subsystem is to process the user query and to search for relevant documents. Firstly, user puts a query. User subsystem will processes it and will assign a keyword to it. Afterwards, the indexation subsystem indexes the query, which will be compared with the document index. "The administrator subsystem

[10]Skovajsová L. Text document retrieval by feed-forward neural networks. Information Sciences and Technologies Bulletin of the ACM Slovakia, 2(2):70-78, 2010.

retrieves relevant documents and sends them to user according to this comparison. The user can use feedback, in which the user marks the most relevant documents from the set of retrieved documents and consequently sends it as a new query. The system creates a new query from these documents and searches again the document base."[11] The three subsystems of the IRS can be represented [26] as a three layer model.

The complex structure of an IR system can be simplified [26] by substitution of the relations between its subsystems by neural networks: the first neural network solves the relation between the user query and the keywords of documents and the second neural network simulates the relation between the keywords and the relevant documents.

Starting from this motivation, one can build a cascade neural network, consisting of the following two neural networks:

(1) the first neural network, associated to the relation between the user query subsystem and the indexing subsystem consists in a fuzzy feedforward neural network (FNP1 or FNP2),
(2) the second neural network used between the indexing subsystem and the administrator subsystem is the Spreading Activation Neural Network (SANN) and it serves for the text document retrieval.

FNP2 (or FNP1) and SANN can be connected together, so that the output of the first neural network is the input of the second network; by combining these two neural networks it results the cascade neural model.

1.6.1 Keyword Recognition Approach Based on the Fuzzy Multilayer Perceptron

The first neural network (see Fig. 1.16) of the cascade neural model, used between the query subsystem and the indexing subsystem contains three neuron layers:

(a) at the input of the first layer, the formulated query by the user is applied; the number of the neurons in the input layer is equal to n, each neuron representing a character of the query;
(b) the hidden layer contains L neurons and it expresses the internal representation of the query;
(c) the third layer (the output layer) contains M neurons, each of them symbolizing a keyword.

The user poses a query on the input and the neuron in the output represents the recognized keyword. A query q can be represented in the same way as a document in the document collection or in a different way.

---

[11] Mokriš, I., and Skovajsová, L., Neural Network Model of System for Information Retrieval from Text Documents in Slovak Language, Acta Electrotechnica et Informatica, 2005, 3(5), 1–6.

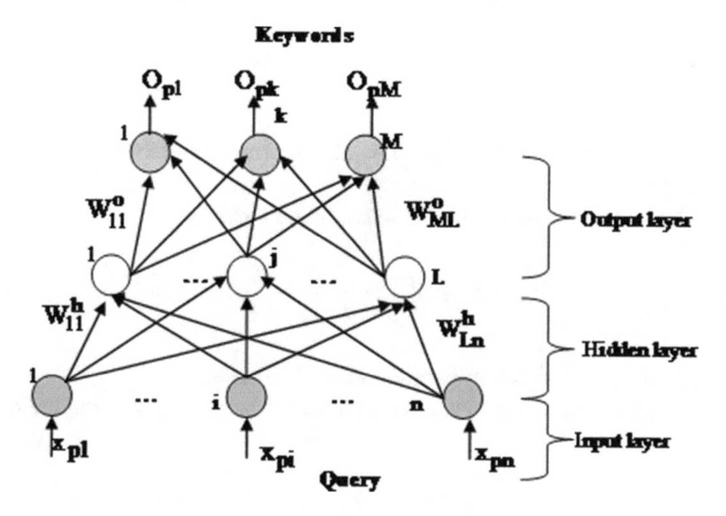

Fig. 1.16 Architecture of FNP2 for determination of keywords

In the case of FNP2, each query q_j, $j = \overline{1, K}$ is represented with a term vector $q_j = (u_{1j}, u_{2j}, \ldots, u_{nj})$ where each weight $u_{ij} \in [0, 1]$ can be computed in the following way:

$$u_{ij} = \frac{f_{ij}}{emax}, \ (\forall)\ i = \overline{1, n},\ j = \overline{1, K}, \tag{1.64}$$

where:

- *emax* represents the maximum number of characters of the query;
- f_{ij} means the frequency of a character (namely, its number of appearances) in a query.

At the input of the FNP1, each component of a query is represented by five values that constitute the membership of the respective component to the five linguistic terms: *unimportant*, *rather unimportant*, *moderately important*, *rather important* and *very important*, depicted in Fig. 1.13.

The five values of each component from a query will be achieved by replacing x with u_{ij}, in each of the relations (1.14)–(1.18).

The FNP2 (or FNP1) training is based on the use of a backpropagation algorithm consisting of the following steps [18, 19]:

Step 1 Set $p = 1$. Apply the vector $X_p = (x_{p1}, \ldots, x_{pn})$ representing a query at the FNP2 input, for which we know the output ideal vector $Y_p = (y_{p1}, \ldots, y_{pM})$, which has its components defined in (1.19). Then, randomly generate the set of the weights corresponding to the hidden layer, i.e., $\{W^h_{ji}\}_{j=\overline{1,L},\ i=\overline{1,n}}$ and the weights related to the output layer: $\{W^o_{kj}\}_{k=\overline{1,M},\ j=\overline{1,L}}$.

Step 2 Calculate the activations of neurons in the hidden layer, using the formula:

$$net^h_{pj} = \sum_{i=1}^{n} W^h_{ji} \cdot x_{pi}, \; (\forall)\, j = \overline{1, L}. \tag{1.65}$$

The hidden layer is used for the query representation based on the formula (1.65).

Step 3 Compute the neuron outputs in the hidden layer:

$$i_{pj} = f^h_j(net^h_{pj}) = \frac{1}{1 + e^{-net^h_{pj}}}, \; (\forall)\, j = \overline{1, L}, \tag{1.66}$$

f^h_j being the activation function in the hidden layer.

Step 4 Calculate the activations of neurons in the output layer, using the relation:

$$net^o_{pk} = \sum_{j=1}^{L} W^o_{kj} \cdot i_{pj}, \; (\forall)\, k = \overline{1, M}. \tag{1.67}$$

Step 5 Compute the neuron outputs in the output layer:

$$O_{pk} = f^o_k(net^o_{pk}) = \frac{1}{1 + e^{-net^o_{pk}}}, \; (\forall)\, k = \overline{1, M}, \tag{1.68}$$

f^o_k being the activation function in the output layer.

Step 6 Refine the weights corresponding to the output layer based on the relation:

$$W^o_{kj}(t+1) = W^o_{kj}(t) + \eta(y_{pk} - O_{pk})O_{pk}(1 - O_{pk}), \; (\forall)\, k = \overline{1, M}, \; j = \overline{1, L}, \tag{1.69}$$

η, $0 < \eta < 1$ meaning the learning rate.

Step 7 Adjust the weights of the hidden layer using the formula:

$$W^h_{ji}(t+1) = W^h_{ji}(t) + \eta\, i_{pj}(1 - i_{pj}) \left(\sum_{k=1}^{M} (y_{pk} - O_{pk})O_{pk}(1 - O_{pk}) W^o_{kj}(t) \right) x_{pi}, \tag{1.70}$$

$(\forall)\, j = \overline{1, L}, \; i = \overline{1, n}$.

Step 8 Compute the error due to the p-th training vector, expressed by the formula:

$$E_p = \frac{1}{2} \sum_{k=1}^{M} (y_{pk} - O_{pk})^2. \tag{1.71}$$

Step 9 If $p < K$ then set $p = p + 1$, namely insert a new vector to the network input. Otherwise, compute the error corresponding to the respective epoch (an epoch means the going through the whole lot of vectors) using the formula:

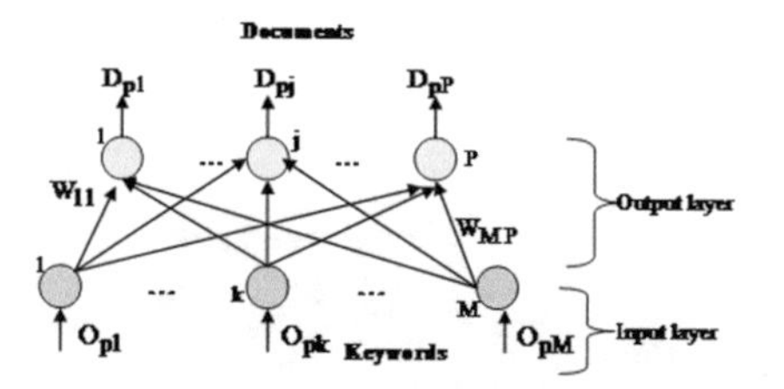

Fig. 1.17 Architecture of SANN for determination of relevant documents

$$E = \frac{1}{K}\sum_{p=1}^{K} E_p \tag{1.72}$$

and start a new training epoch.

The training algorithm one finishes after a certain number (fixed) of epochs.

1.6.2 Text Document Retrieval Approach on the Base of a Spreading Activation Neural Network

Spreading Activation Neural Network (SANN) serves for the text document retrieval, where document collection that contains P documents and M terms (keywords) can be represented by the VSM matrix.[12] The VSM matrix A (the relative frequency matrix of keywords in document set), by the size $M \times P$ has following structure [28]: each column represents a document and each row signifies a term; hence $A(k, j)$ is the frequency of the keyword k in the document j:

$$\mathbf{A} = \begin{array}{c} \\ k_1 \\ k_2 \\ \\ k_M \end{array} \begin{array}{c} \begin{array}{cccc} d_1 & d_2 & & d_P \end{array} \\ \begin{pmatrix} W_{11} & W_{12} & \ldots & W_{1P} \\ W_{21} & W_{22} & \ldots & W_{2P} \\ \ldots & \ldots & \ldots & \ldots \\ W_{M1} & W_{M2} & \ldots & W_{MP} \end{pmatrix} \end{array}.$$

When the document collection is large, we shall apply the DCT for each row of the VSM matrix to reduce the dimension of the document space that represents recognition feature space and to obtain the latent semantic model.

Figure 1.17 depicts the SANN; it has two neuron layers:

(a) the first layer, which contains M neurons, each of them corresponding to a neuron of the output layer of the FNP2 or FNP1; the keyword vector which is applied to the network input represents the query;

[12]Liu, B., Web Data Mining, Springer-Verlag Berlin Heidelberg, 2008.

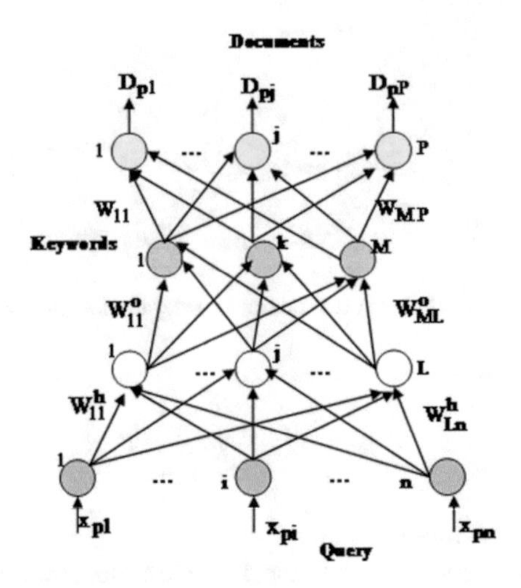

Fig. 1.18 Architecture of cascade neural network model of IR system

(b) the second layer (the output layer), containing P neurons, number which coincides with the number of documents from the database; the output vector symbolizes the similarities between the query formulated by the user and the documents that make up the database.

The neuron of the input layer of SANN is the same as the neuron of the output layer of first neural network defined by (1.68). The related weights to this network, that perform the connection of the neurons from the input layer with neurons from the output layer coincide with the elements of the matrix A, namely $W = A$. SANN hasn't a training algorithm as its weights are determined on the basis of relative frequency matrix. The neuron outputs in the output layer will be:

$$net_{pj} = f_j(net_{pj}) = \sum_{k=1}^{M} W_{kj} \cdot O_{pk}, \ (\forall)\, j = \overline{1, P}. \tag{1.73}$$

On the base of the FNP2 (or FNP1) and SANN, the following cascade neural network can be built (see Fig. 1.18).

"Many university, corporate, and public libraries now use IR systems to provide access to books, journals, and other documents."[13]

"Dictionary and encyclopedia databases are now widely available for PCs. IR has been found useful in such disparate areas as office automation and software engineering. Indeed, any discipline that relies on documents to do its work could

[13]Mihăescu, C., Algorithms for Information Retrieval Introduction, 2013, http://software.ucv.ro/cmihaescu/ro/teaching/AIR/docs/Lab1-Algorithms%20for%20Information%20Retrieval.%20Introduction.pdf.

Fig. 1.19 Importance of using the IR system (see footnote 16)

potentially use and benefit from IR."[14] It is used today in many applications, being "used to search for documents, content thereof, document metadata within traditional relational databases or internet documents more conveniently and decrease work to access information." (see footnote 14). The relevant information needed by a user is based on the retrieved documents, achieved using the search engines as Google, Yahoo or Microsoft Live Search. A lot of problems in IR "can be viewed as a prediction problem, i.e., to predict ranking scores or ratings of web pages, documents, music songs etc. and learning the information desires and interests of users." (see footnote 14).

"Managing the vast amount of online information and classifying it into what could be relevant to our needs is an important step toward being able to use this information."[15]

Figure 1.19 shows how much popular is the application of the Web Classification "not only to the academic needs for continuous knowledge growth, but also to the

---

[14]Mihăescu, C., Algorithms for Information Retrieval Introduction, 2013, http://software.ucv.ro/~cmihaescu/ro/teaching/AIR/docs/Lab1-Algorithms%20for%20Information%20Retrieval.%20Introduction.pdf.

[15]Xhemali, D., and Hinde, C.J., and Stone, R.G., Naïve Bayes vs. Decision Trees vs. Neural Networks in the Classification of Training Web Pages, International Journal of Computer Science Issues, 2009, 4(1), 16–23.

needs of industry for quick, efficient solutions to information gathering and analysis in maintaining up-to-date information that is critical to the business success."[16]

References

1. L. Chen, D. H. Cooley, and J. Zhang. Possibility-based fuzzy neural networks and their application to image processing. *IEEE Transactions on Systems, Man, and Cybernetics*, 29(1):119–126, 1999.
2. B. Hammer and T. Villmann. Mathematical aspects of neural networks. In *11th European Symposium on Artificial Neural Networks (ESANN' 2003)*, 2003.
3. T. Hastie and ands Friedman J. Tibshirani, R. *The Elements of Statistical Learning. Data Mining, Inference, and Prediction*. Springer-Verlagn Berlin Heidelberg, 2009.
4. M.A. Razi and K. Athappilly. A comparative predictive analysis of neural networks (NNs), nonlinear regression and classification and regression tree (cart) models. *Expert Systems with Applications*, 29:65–74, 2005.
5. V. Reshadat and M.R. Feizi-Derakhshi. Neural network-based methods in information retrieval. *American Journal of Scientific Research*, 58:33–43, 2012.
6. B. Zaka. Theory and applications of similarity detection techniques. http://www.iicm.tugraz.at/thesis/bilal_dissertation.pdf, 2009.
7. B.M. Ramageri. Data mining techniques and applications. *Indian Journal of Computer Science and Engineering*, 1(4):301–305, 2010.
8. H. Bashiri. Neural networks for information retrieval. http://www.powershow.com/view1/1af079-ZDc1Z/Neural_Networks_for_Information_Retrieval_powerpoint_ppt_presentation, 2005.
9. J. Mehrad and S. Koleini. Using som neural network in text information retrieval. *Iranian Journal of information Science and Technology*, 5(1):53–64, 2007.
10. K.A. Olkiewicz and U. Markowska-Kaczmar. Emotion-based image retrieval an artificial neural network approach. In *Proceedings of the International Multiconference on Computer Science and Information Technology*, pages 89–96, 2010.
11. I. Iatan and M. de Rijke. Mathematical aspects of using neural approaches for information retrieval. *Complex and Intelligent Systems (Reviewers Assigned)*, 2016.
12. A.N. Netravali and B.G. Haskell. *Digital Pictures: Representation and Compression*. Springer, 2012.
13. R.C. Gonzales and A. Woods. *Digital Image Processing*. Prentice Hall, second edition, 2002.
14. W. Burgerr and M.J. Burge. *Principles of Digital Image Processing. Fundamental Techniques*. Springer-Verlag London, 2009.
15. A. Vlaicu. *Digital Image Processing (in Romanian)*. MicroInformatica Group, Cluj-Napoca, 1997.
16. V.E. Neagoe. Pattern recognition and artificial intelligence (in Romanian), lecture notes, Faculty of Electronics, Telecommunications and Information Technology, University Politehnica of Bucharest. 2000.
17. M. Ettaouil, Y. Ghanou, K. El Moutaouakil, and M. Lazaar. Image medical compression by a new architecture optimization model for the Kohonen networks. *International Journal of Computer Theory and Engineering*, 3(2):204–210, 2011.
18. V. E. Neagoe and O. Stănăşilă. *Pattern Recognition and Neural Networks (in Romanian)*. Ed. Matrix Rom, Bucharest, 1999.
19. I. Iatan. *Neuro-Fuzzy Systems for Pattern Recognition (in Romanian)*. PhD thesis, Faculty of Electronics, Telecommunications and Information Technology-University Politehnica of Bucharest, PhD supervisor: Prof. dr. Victor Neagoe, 2003.

[16]Liu, T. Y., Learning to Rank for Information Retrieval, 2011, Springer-Verlag Berlin Heidelberg.

20. L.T. Huang, L.F. Lai, and C.C Wu. A fuzzy query method based on human-readable rules for predicting protein stability changes. *The Open Structural Biology Journal*, 3:143–148, 2009.
21. A. Ganivada and S.K. Pal. A novel fuzzy rough granular neural network for classification. *International Journal of Computational Intelligence Systems*, 4(5):1042–1051, 2011.
22. Q. Ni, C. Guo, and J. Yang. Research of face image recognition based on probabilistic neural networks. In *IEEE Control and Decision Conference*, 2012.
23. Y. Sun, X. Lin, and Q. Jia. Information retrieval for probabilistic pattern matching based on neural network. In *International Conference on Systems and Informatics, ICSAI2012*, 2012.
24. G.A. Anastassiou and I. Iatan. A new approach of a possibility function based neural network. In *Intelligent Mathematics II: Applied Mathematics and Approximation Theory*, pages 139–150. Springer International Publishing, 2016.
25. L. Skovajsová. Text document retrieval by feed-forward neural networks. *Information Sciences and Technologies Bulletin of the ACM Slovakia*, 2(2):70–78, 2010.
26. I. Mokriš and L. Skovajsová. Neural network model of system for information retrieval from text documents in slovak language. *Acta Electrotechnica et Informatica*, 3(5):1–6, 2005.
27. T.N. Yap. Automatic text archiving and retrieval systems using self-organizing kohonen map. In *Natural Language Processing Research Symposium*, pages 20–24, 2004.
28. B. Liu. *Web Data Mining*. Springer-Verlag Berlin Heidelberg, 2008.

Chapter 2
A Fuzzy Kwan–Cai Neural Network for Determining Image Similarity and for the Face Recognition

2.1 Introduction

Similarity is a crucial issue in Image Retrieval [1–4]. It is relevant both for unsupervised clustering [5, 6] and for supervised classification [7]. In this study, we aim to provide an effective method for learning image similarity. To reach this aim we start of from the Fuzzy Kwan–Cai Neural Network (FKCNN) [8] and turn it into a supervised method for learning similarity. Unlike the classical unsupervised FKCNN [8], in which a class is represented by a single output neuron, the supervised FKCNN has more output neurons that each designate a class. These give a better performance than their unsupervised counterparts as in the case of the classical unsupervised FKCNN, a class is represented by a single output neuron while the supervised FKCNN has more output neurons that each designate a class. This concept is similar with the idea of replacing the binary decision about the membership (nonmembership) of a pattern to a class, with the introduction of a membership degree, between 0 and 1.

In order to evaluate the performance of our proposed neural network, it is compared with two baseline methods: Self-organizing Kohonen maps (SOKM) and k-Nearest Neighbors (k-NN).

The feasibility of the presented methods for similarity learning has been successfully evaluated on the Visual Object Classes (VOC) database [9], that consists of 10102 images in 20 object classes.

The concept of similarity is important not only in almost every scientific field but it has [10, 11] deep roots in philosophy and psychology. Our work however deals more with the measure of similarity in computer science domain (Information Retrieval to be more specific, that has focused on images, video, and to some extent audio).

"Measuring image similarity is an important task for various multimedia applications."[1]

---

[1] Perkiö, J., and Tuominen, A., and Myllymäki, P., Image Similarity: From Syntax to Weak Semantics using Multimodal Features with Application to Multimedia Retrieval, Multimedia Information Networking and Security, 2009, 1, 213–219.

I.F. Iatan, *Issues in the Use of Neural Networks in Information Retrieval*,
Studies in Computational Intelligence 661, DOI 10.1007/978-3-319-43871-9_2

Use of adequate measures improves [10] the accuracy of information selection. As there are a lot of ways to compute similarity or dissimilarity among various object representations, they need to be categorized [10] as:

(1) *distance-based similarity measures*
Example of this approach include following models/methods: Minkowski distance, Manhattan/City block distance, Euclidean distance, Mahalanobis distance, Chebyshev distance, Jaccard distance, Dice's coefficient, cosine similarity, Hamming distance, Levenshtein distance, soundex distance.

(2) *feature-based similarity measures (contrast model)*
This method (suggested by Tversky in 1977) constitutes an alternate to distance-based similarity measures, where the similarity is computed by common features of compared entities. The entities are more similar if they share more common features and dissimilar in case of more distinctive features. The following formula can be used to determine similarity between entities A and B:

$$s(A, B) = \alpha g(A \cap B) - \beta g(A - B) - \gamma g(B - A), \tag{2.1}$$

where α, β, γ are used to determine the respective weights of associated values, $g(A \cap B)$ represents the common features in A and B, $g(A - B)$ represents distinctive features of A and $g(B - A)$ that of entity B.

(3) *probabilistic similarity measures*
In order to calculate relevance among some complex data types, use of the following probabilistic similarity measure are required: maximum likelihood estimation, maximum a posteriori estimation.

(4) *extended/additional measures*: similarity measures based on fuzzy set theory [12], similarity measures based on graph theory, similarity-based weighted nearest neighbors [13, 14], similarity-based neural networks [11].

Next to the definition of similarity measures [15], "in the last few decades, the perception of similarity received a growing attention from psychological researchers."[2]

As the investigation of similarity is crucial for many classification methods, more recently, learning similarity measure [16] has also attracted attention in the machine learning community [17, 18].

"The problem of classifying samples based only on their pairwise similarities can be divided into two subproblems: measuring the similarity between samples and classifying the samples based on their pairwise similarities."[3]

[2]Melacci, S., and Sarti, L., and Maggini, M. and Bianchini, M., A Neural Network Approach to Similarity Learning, ANNPR 2008, LNAI 5064, 133–136.

[3]Cazzanti, L., Generative Models for Similarity-based Classification, 2007, http://www.mayagupta.org/publications/cazzanti_dissertation.pdf.

"Similarity-based classifiers estimate the class label of a test sample based on the similarities between the test sample and a set of labeled training samples, and the pairwise similarities among the training samples."[4]

Similarity-based classification is useful for problems in Multimedia Analysis [19, 20], Computer Vision, Bioinformatics, Information Retrieval [21–23], and a broad range of other fields."Among the various traditional approaches of pattern recognition [24], the statistical approach has been most intensively studied and used in practice. More recently, the addition of artificial neural network techniques theory have been receiving significant attention. The design of a recognition system requires careful attention to the following issues: definition of pattern classes, sensing environment, pattern representation, feature extraction and selection, cluster analysis, classifier design and learning, selection of training and test samples, and performance evaluation."[5]

NNs methods hold great promise for defining similarity.

In this chapter [1], two similarity learning approaches based on Artificial Neural Networks (ANNs) are presented. We have been very motivated to write the present paper to highlight the fact that neural networks outperform statistical techniques in computing similarities, too. The main contributions of this chapter are:

- we provide a neural method for learning image similarity, by turning the classical unsupervised Fuzzy Kwan–Cai Neural Network (FKCNN) into a supervised method, which gives a better performance than its unsupervised counterparts;
- we build a novel algorithm based on an evaluation criteria to compare the performances of the three presented methods;
- we test the resulting similarity functions on the PASCAL Visual Object Classes (VOC) data set, which consists in 20 object classes;
- we perform a comparative study of the proposed similarity learning method and compare it with two baseline methods: Self-organizing Kohonen Maps (SOKM) and k-Nearest Neighbor rule (k-NN) in terms of their ability to learn similarity; these particular algorithms (k-NN and SOKM) have been chosen as benchmark for performance comparison of the improved FKCNN algorithm because of the fact that all of them (k-NN, which is a nonneural method, SOKM—an unsupervised neural network, FKCNN—a supervised neural network, resulted by improving the unsupervised FKCNN) are some *extended/additional measures* to compute similarity or dissimilarity among various object representations.
- we highlight the overall performance of FKCNN, being better for our task than SOKM and k-NN.

---

[4]Chen, Y., and Garcia, E.K., and Gupta, M.Y., and Rahimi, A., and Cazzanti, A., Similarity-based Classification: Concepts and Algorithms, Journal of Machine Learning Research, 2009, 10, 747–776.

[5]Basu, J.K., and Bhattacharyya, D., and Kim, T.H., Use of Artificial Neural Network in Pattern Recognition, International Journal of Software Engineering and Its Applications, 2010, 4(2), 22–34.

The ANNs are configured [25] for a specific application, such as pattern recognition [26] or data classification, through a learning process. "The advantages for using an ANN in similarity matching are twofold: one that we can combine multiple features extracted with different methods, and the second that the combination of these features can be nonlinear."[6]

2.2 Related Work

In past few years, many papers have been proposed to efficiently and accurately estimate the similarity. Yu et al. (2012) provided [27] a novel semisupervised multiview distance metric learning (SSM-DML), which learns the multiview distance metrics both from the multiple feature sets and from the labels of unlabeled cartoon characters, under the umbrella of graph-based semisupervised learning. The effectiveness of the SSM-DML has been proven in the cartoon applications.

In the same year, Yu et al. (2012) achieved [28] a novel method which learns novel distance through hypergraph for transductive image classification. Hypergraph learning has been projected to solve the following difficulties: the existing graph-based learning methods can only model the pairwise relationship of images and they are sensitive to the parameter used in similarity calculation. In their proposed method, the authors generate hyperedges by linking images and their nearest neighbors and can automatically modulate the effects of different hyperedges (by learning the labels of unlabeled images and the weights of hyperedges).

Later, Yu et al. (2014) proposed [29] a novel multiview stochastic distance metric learning method, which is crucial in utilizing the complementary information of multiview data. In comparison with the existing strategies, the proposed approach adopts the high-order distance obtained from the hypergraph to replace pairwise distance in estimating the probability matrix of data distribution.

Kwan and Cai, 1997 have introduced [8] a four layer unsupervised fuzzy neural network to solve pattern recognition problems, proving the importance of combining of strengths of fuzzy logic and neural networks. The fuzzy neurons of the third layer allow the computing of the similarities corresponding to an input pattern to all the learned patterns. In the case of the classical unsupervised FKCNN, a class is represented by a single output neuron.

Hariri et al. (2008) modified [7] the structure of the unsupervised fuzzy neural network of Kwan and Cai, composing a new version (improved five-layer feedforward Supervised Fuzzy Neural Network = SFN) and used it for classification and identification of shifted and distorted training patterns. To show the identification capability of the SFN, they used fingerprint pattern to prove that their testing result is more considerable than early FKCNN.

[6]Chen, F., Similarity Analysis of Video Sequences Using an Artificial Neural Network, University of Georgia, 2003, http://athenaeum.libs.uga.edu/bitstream/handle/10724/6674/chen_feng_200305_ms.pdf?sequence=1.

We started this work with the paper [30], which presents a neuro-fuzzy approach to face recognition using an improved version of Kwan and Cai fuzzy neural network [8]. We have modified the above-mentioned fuzzy net from an unsupervised network into a supervised one, called *Fuzzy Kwan–Cai Supervised Net* and we have applied it for the special task of *Face Recognition*. The supervised FKCNN has more output neurons that each designate a class. Classification of an image with unknown class is achieved through the association of the input pattern with the class corresponding to that neuron in the fourth layer which has the output equal to 1.

2.3 Background

2.3.1 Measure of the Class Similarities

Let us first carefully examine the notion of similarity, before defining the methods for computing similarity. We shall use the notion of class similarities as it will be necessary later, in the algorithm of the evaluation criteria. Let us suppose we have M classes ω_i, $i = \overline{1, M}$, with mean vectors μ_i and inner variances S_i. We denote with D_{ij} the distance between the feature vectors of the classes ω_i, ω_j; usually one chooses μ_i as a feature vector of the class ω_i. We have a measure $R(S_i, S_j, D_{ij})$ of class similarities [31] if for each $i, j = \overline{1, M}$ it defines the *similarity between the classes* ω_i and ω_j such that:

(1) $R(S_i, S_j, D_{ij}) \geq 0$ (the similarity measure among the classes is greater than or equal to zero);
(2) $R(S_i, S_j, D_{ij}) = R(S_i, S_j, D_{ji})$, namely the similarity measure between two pattern classes is necessarily symmetric;
(3) $R(S_i, S_j, D_{ij}) = 0$ if and only if $S_i = S_j = 0$ (the similarity measure is null if and only if the inner variances are null);
(4) If $S_j = S_k$ and $D_{ij} < D_{ik}$, then $R(S_i, S_j, D_{ij}) < R(S_i, S_k, D_{ik})$, namely the similarity measure decreases as a consequence of the fact that the inner variances are the same and the distance between the classes, i.e., between the feature vectors increases;
(5) If $D_{ij} = D_{ik}$ and $S_j > S_k$, then $R(S_i, S_j, D_{ij}) > R(S_i, S_k, D_{ik})$, i.e., the similarity measure increases in the case when the distances among the classes are equal and the inner variances increase.

Table 2.1 The important notations of this subsection

The used notation	Its significance
M	Number of the classes $\omega_i,\ i=\overline{1,M}$
$R(S_i, S_j, D_{ij})$	A measure of class similarities
μ_i	Mean of the class ω_i
S_i	Inner variances of the samples belonging to the class ω_i
D_{ij}	Minkowski distance of order p between μ_i and μ_j;
	For $p=2$ it means the Euclidean distance

An example of a similarity measure which we shall use throughout this paper is [31]:

$$R_{ij} = R(S_i, S_j, D_{ij}) = \frac{S_i + S_j}{D_{ij}},\ (\forall)\ i,j = \overline{1,M}. \tag{2.2}$$

$$\mu_i = \frac{1}{N_i} \sum_{X_j \in \omega_i} X_j, \tag{2.3}$$

$$S_i = \left(\frac{1}{N_i} \sum_{X_j \in \omega_i} \| X_j - \mu_i \|^q \right)^{1/q}, \tag{2.4}$$

$$D_{ij} = \left(\sum_{k=1}^{n} |\mu_{ik} - \mu_{jk}|^p \right)^{1/p}, \tag{2.5}$$

D_{ij} being the Minkowski distance of order p between μ_i and μ_j; for $p = 2$ it will result the Euclidean distance. For $q = 2$, S_i represents the inner variances of the samples belonging to the class ω_i against the mean μ_i of that class. If $p = q = 2$ we get the similarity measure introduced by Fisher [31].

We shall provide a notation Table 2.1 to list the important notations of this subsection.

2.3.2 Similarity-Based Classification

"Similarity-based classifiers are defined as those classifiers that require only a pair-wise similarity—a description of the samples themselves is not needed."[7]

[7] Mellouk, A., and Chebira, A., Machine Learning, 2009, InTech.

A simple similarity-based classifier is the k-Nearest Neighbor classification rule (k-NN), classifying an image into the class of the most similar images in the database [32]. The probability of error for the k-NN rule is less than the probability of error corresponding to the NN. The NN rule is suboptimal, i.e., it does not converge to Bayes optimal rule, but it turns out [33] that the probability of error of NN rule, asymptotically never exceeds twice the Bayes rate. Therefore, the probability of error for NN rule has the lower bound and the upper bound expressed in the formula [33]:

$$P^* \leq P \leq P^* \left(2 - \frac{c}{c-1} P^*\right),$$

where c is the number of classes and P^* denote the error rate of Bayes rule.

Self-organizing Kohonen maps are not designed for similarity-based classification. This neural network has, however, proved to be very useful in a wide range of problems, and it is considered especially suitable for clustering large high-dimensional data sets such as images, documents [34, 35], etc. It is unsupervised, in that it does not need any categorical information about the input data during its training process. After the training process, the achieved map is organized in such a way that similar data from the input space are mapped to the same unit or to neighboring units on the map.

The Kohonen neural network is a network whose learning process is based on [31]:

- *the principle of competition*: for a vector applied to the network input, one determines the winner neuron in the network, as the neuron with the best match;
- *the neighborhood principle*: one refines both the weight vector associated with the winner neuron and the weights associated with the surrounding neurons.

Figure 2.1 proves that the SOKM transforms the similarities among the input vectors into some neighborhood relationships between the neurons; it can be noticed that three similar input vectors from the class 1 are assigned to three neighboring neurons, etc.

The SOKM is characterized in that the neighboring neurons:

- are correlated;
- turn into some specific detectors of different classes of patterns;
- characterize the vectors applied to the network input.

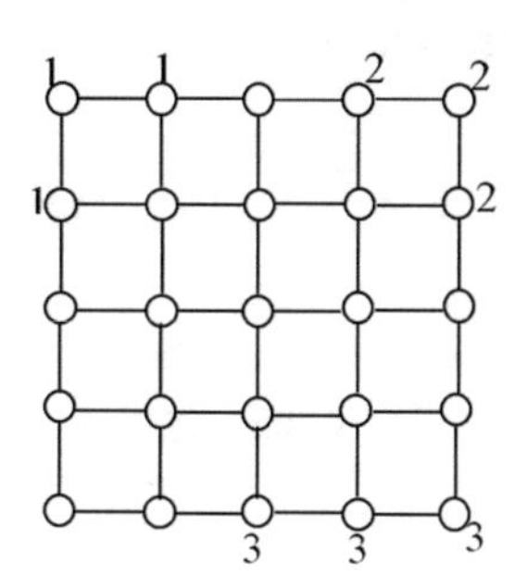

Fig. 2.1 Similarities among the input vectors

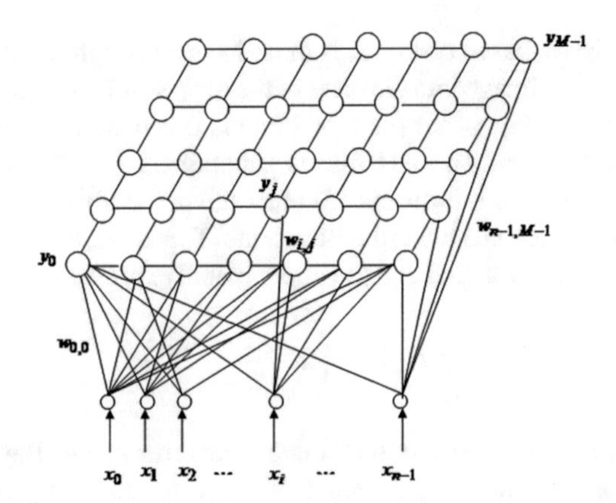

Fig. 2.2 Structure of a rectangular Kohonen network with M output neurons and n input neurons

The Kohonen network structure has two layers of neurons (see Fig. 2.2).

The neurons of the input layer correspond to the dimension of the input vector. The output layer contains M neurons, this number being greater than or equal to the number of clusters. These neurons are constrained to a regular grid, which usually is two-dimensional. All neurons in the input layer are connected to all neurons in the output layer. This neural network is self-organizing. The maps are trained in an unsupervised way using a competitive learning scheme.

The network training is one competitive, unsupervised, and self-organizing.

The training algorithm corresponding to the SOKM can be summarized below:

Step 1 Set t = 0. Initialize the learning rate $\eta(0)$ and the neighborhood of the neuron j, in the frame of the network, denoted V_j and establish the rule regarding their variations in time.
The weight values of the SOKM are initialized with some random values. Each neuron j of SOKM is characterized by a vector of weights $W_j = (w_{0j}, w_{1j}, \ldots, w_{n-1,j})^t \subset \Re^n$, $(\forall)\, j = \overline{0, M-1}$, M being the number of neurons of SOKM and n representing the dimension of the input vectors.

Step 2 Apply the vector $X_p = (x_{p0}, x_{p1}, \ldots, x_{p,n-1})^t \subset \Re^n$, $(\forall)\, p = \overline{1, N}$ (N being the number of the training vectors) at the network input.

Step 3 Compute the Euclidean distances between X_p and the weight vectors associated with the network neurons, at the time moment t, based on the formula:

$$d_j = \| X_p(t) - W_j(t) \|^2 = \sum_{i=0}^{n-1} (x_{pi}(t) - w_{ij}(t))^2, \; (\forall)\, j = \overline{0, M-1}. \quad (2.6)$$

Step 4 Find the winning neuron j^* of the SOKM, by computing the minimum distance among the vector X_p and the weight vectors W_j of the network, according to the relation [31]:

$$d_{j^*} = \min_{j=0}^{M-1} d_j, \ (\forall)\, j = \overline{0, M-1}. \tag{2.7}$$

Step 5 Once the winning neuron is known, one refines the weights for the winning neuron and its neighbors, using the relation [31]:

$$W_{j+1}(t) = W_j(t) + \eta(t) \cdot \left(X_p(t) - W_j(t)\right), \ j \in V_{j^*}, (\forall)\, i = \overline{0, n-1}, \tag{2.8}$$

where

$$\eta(t) = \frac{\eta_0}{t^p} \exp\left(-\frac{\|r_k - r_{j^*}\|}{\sigma^2}\right) \tag{2.9}$$

is the learning rate and V_{j^*} represents the neighborhood of the neuron j^*.

In the relation (2.9) we have:

- $\eta_0 = \eta_0(t)$ and $\sigma = \sigma(t)$ is the value of the learning rate in the neighborhood center and respectively the parameter that controls the speed of the decrease of the learning rate depending on the neighborhood radius;
- r_{j^*} and r_k are the position vectors within the network of the central neuron of the neighborhood V_{j^*} and respectively of the neuron k for which we update the weights.

The neighborhood radius V_{j^*} can vary in time, being chosen to the beginning of refining to cover the whole network, after which it decreases monotonically in time.

Step 6 Set p = p + 1. If we finished the whole lot of the vectors, we have to check if the stop condition of the training process is satisfied, namely: after a fixed number of epochs or when we no longer get a significant change of the weights associated with the network neurons, according to [31]:

$$\| w_{ij}(t+1) - w_{ij}(t) \| < \varepsilon, (\forall)\, i = \overline{0, n-1}, (\forall)\, j = \overline{0, M-1}. \tag{2.10}$$

If the stop condition from (2.10) is not satisfied then we resume the training algorithm.

When we want to classify the lot of input vectors based on the similarities between the vectors, it will be made a training of the network until all the vectors belonging to a class will be associated to a single neuron of the network, which will represent that class.

The performances of SOKM depend on the network shape; it can be with the following architecture:

(1) 1D: linear, circular;
(2) 2D: planar (rectangular, hexagonal), spherical, toroidal;
(3) 3D: spherical, toroidal, parallelepipedal.

As the similarity is not relevant only for the unsupervised clustering but for the supervised classification too, we shall turn the classical FKCNN, with four types of fuzzy neurons, into a supervised one.

When an input pattern is provided to the FKCNN input, this network first fuzzifies the respective pattern and then computes the similarities of this pattern to all of the learnt patterns.

In the case when a learnt previously pattern one presents at the FKCNN input, the network will treat the respective pattern as a known pattern, without to relearn it.

After computing the similarities, the FKCNN takes a decision by selecting the learnt pattern with the highest similarity and gives a nonfuzzy output.

2.3.3 Using Kohonen Algorithm in Vector Quantization

In the previous section, we have seen that the SOKM is a unsupervised method for learning similarity; SOKM also constitutes a nonlinear neural approach of feature selection (see Sect. 2.6.2). The aim of the present section is to describe how the SOKM can perform a Vector Quantization (VQ).

VQ has been noticed [36–39] as an efficient technique for image compression.

The VQ compression system consists of the two components [37, 40]:

(1) VQ encoder;
(2) VQ decoder.

New approaches based on neural networks as compression tools have been dealt by [41–43]. Particularly, the SOKM can be used as a vector quantizer for images [36–38].

We shall describe the algorithm which performs the VQ for the 256×256 images, using the SOKM [44]:

Step 1 Split the original image into 1024 square blocks 8×8 and apply the DCT for each of them as in the case of the Algorithm 1.1.

Step 2 Build a vector, whose components are the nine coefficients (10 retained without the first one) from each block, in a zigzag fashion (see the Algorithm 1.1) and cancel the rest of $(64 - 10)$ coefficients; in this way, it results the set of the nine-dimensional vectors X_i, $i = \overline{1, 1024}$, denoted $\aleph = \{X_1, \ldots, X_{1024}\}$.

Step 3 Design a SOKM with $32 \times 32 = 1024$ output neurons, which will be trained using the 1024 vectors achieved at the *Step 2*, that will be applicated one at a time, at the input network.

Step 4 "Freeze" the weights resulted at the end of the learning algorithm of the SOKM.

Step 5 Determine the prototype vectors for each neuron in the following way:

(a) apply again each vector X_i, $i = \overline{1, 1024}$ at the input of the SOKM to compute for its corresponding Euclidean distance to each neuron from the network with the formula (2.6);

(b) find the minimum distance among the vector X_i, $i = \overline{1, 1024}$ and the weight vectors, that characterize each neuron of the SOKM, according to the relation (2.7);

(c) design a statistics in order to determine how many vectors (each of them representing one block) from $\aleph$ are associated to one neuron and then calculate the mean of these vectors to find the prototype vectors for each neuron.

Remark 2.1 Each class is characterized by a:

1. neuron j^*;
2. prototype vector μ_{j^*} attached to this neuron;
3. 10-bit code word $C_{j_*} = (c_{1j_*}, \ldots, c_{10,j_*})$ (it corresponds to a network with 1024 output neurons); for example $C_{j^*} = \underbrace{(11\ldots0)}_{10\ times}$

Step 6 Build the coded image by replacing (for each of the 1024 blocks) the first from the 10 retained coefficients with c_{1j_*} (the first bit of the code word, which represents the respective block).

Step 7 Apply the inverse DCT for each of the 1024 blocks to achieve the compressed image and convert its elements into integer values.

Step 8 Evaluate the performances of the VQ algorithm both from the visual point view and in terms of the PSNR, too.

2.3.4 Fourier Descriptors

The descriptors of different objects can be compared in order to give [45] a measurement of their similarity.

The Fourier descriptors have interesting properties in describing the shape of an object, by considering its boundaries.

Let γ be [31] a closed pattern, oriented counterclockwise, having the parametric representation: $z(l) = (x(l), y(l))$, where l is the length of an arc in a circle, along the curve γ, starting from an origin and $0 \le l < L$, where L is the length of the boundary.

A point moving along the boundary generates the complex function $u(l) = x(l) + iy(l)$, which is a periodic function, having the period L.

Definition 2.1 The *Fourier descriptors* represents the coefficients corresponding to the decomposition of the function $u(l)$ in a complex Fourier series.

Similar to the implementation of a Fourier series to build a specific time signal consisting of cosine/sine waves of different amplitude and frequency, "the Fourier descriptor method uses a series of circles with different sizes and frequencies to build up at two-dimensional plot of a boundary"[8] belonging to an object.

[8]Janse van Rensburg, F.J., and Treurnicht, J., and Fourie, C.J., The Use of Fourier Descriptors for Object Recognition in Robotic Assembly, 5th CIRP International Seminar on Intelligent Computation in Manufacturing Engineering, 2006.

The Fourier descriptors are computed using the formula [31, 46]:

$$a_n = \frac{1}{L}\int_0^L u(l)\mathrm{e}^{-\mathrm{i}\frac{2\pi}{L}nl}\mathrm{d}l \tag{2.11}$$

such that

$$u(l) = \sum_{n=-\infty}^{\infty} a_n \cdot \mathrm{e}^{\mathrm{i}\frac{2\pi}{L}nl}. \tag{2.12}$$

We shall prove what becomes the formula (2.11) in the case of a polygonal contour, depicted in Fig. 2.3.

Denoting

$$\lambda = \frac{\|\overline{V_{k-1}M}\|}{\|\overline{MV_k}\|}, \tag{2.13}$$

namely (see Fig. 2.4):

$$\lambda = \frac{l - l_{k-1}}{l_k - l}, \tag{2.14}$$

where $M(x_M, y_M)$, $V_{k-1}(x_{k-1}, y_{k-1})$, $V_k(x_k, y_k)$ and

$$\begin{cases} l_0 = 0, \\ l_k = \sum_{i=1}^{k} |V_i - V_{i-1}|; \end{cases} \tag{2.15}$$

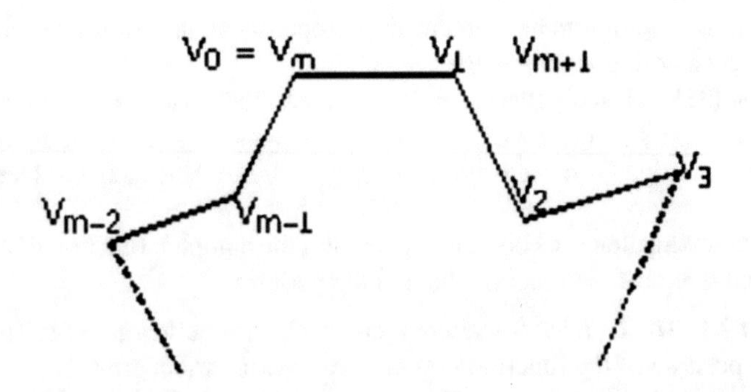

Fig. 2.3 Polygonal contour

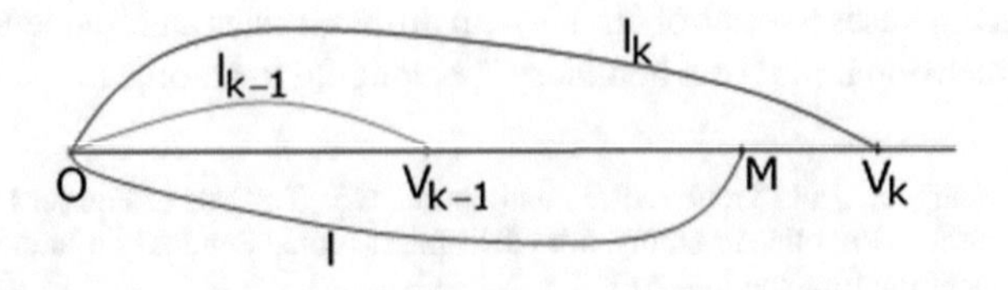

Fig. 2.4 Parametric representation of a segment

from (2.11) it results:

$$a_n = \frac{1}{L}\sum_{k=1}^{m}\int_{l_{k-1}}^{l_k}(x_M(l) + \mathrm{i}y_M(l))\cdot \mathrm{e}^{-\mathrm{i}\frac{2\pi}{L}nl}\mathrm{d}l. \tag{2.16}$$

From (2.13) we deduce:

$$\begin{cases} x_M = \frac{x_{k-1}+\lambda x_k}{1+\lambda}, \\ y_M = \frac{y_{k-1}+\lambda y_k}{1+\lambda}. \end{cases} \tag{2.17}$$

We shall treat each coordinate pair as a complex number so that in Fig. 2.5:

Taking into account the previous assumption and the relations (2.17) and (2.14) we can notice that

$$x_M(l) + \mathrm{i}y_M(l) = \frac{x_{k-1} + \frac{l-l_{k-1}}{l_k-l}\cdot x_k}{1+\frac{l-l_{k-1}}{l_k-l}} + \mathrm{i}\frac{y_{k-1} + \frac{l-l_{k-1}}{l_k-l}\cdot y_k}{1+\frac{l-l_{k-1}}{l_k-l}}$$

$$= \frac{(x_{k-1}+\mathrm{i}y_{k-1})l_k + l[(x_k - x_{k-1}) + \mathrm{i}(y_k - y_{k-1})] - (x_k + \mathrm{i}y_k)l_{k-1}}{l_k - l_{k-1}}$$

$$= \frac{V_{k-1}l_k + l(V_k - V_{k-1}) - V_k l_{k-1}}{l_k - l_{k-1}};$$

hence, the formula (2.16) will become

$$a_n = \frac{1}{L}\sum_{k=1}^{m}\frac{1}{l_k - l_{k-1}}\int_{l_{k-1}}^{l_k}[V_{k-1}l_k + l(V_k - V_{k-1}) - V_k l_{k-1}]\cdot \mathrm{e}^{-\mathrm{i}\frac{2\pi}{L}nl}\mathrm{d}l$$

$$= \frac{1}{L}\sum_{k=1}^{m}\frac{1}{l_k - l_{k-1}}\int_{l_{k-1}}^{l_k}[V_{k-1}l_k + l(V_k - V_{k-1}) - V_k l_{k-1}]\cdot \mathrm{e}^{-\mathrm{i}\frac{2\pi}{L}nl}\mathrm{d}l =$$

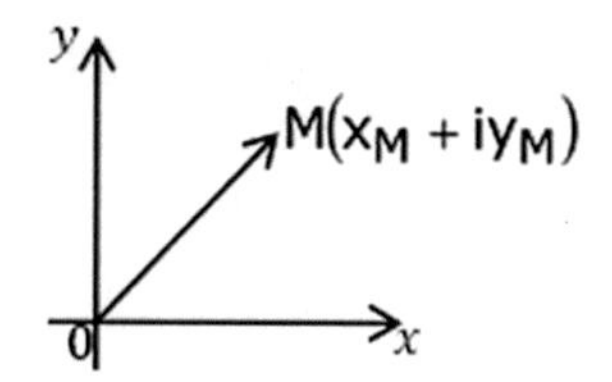

Fig. 2.5 Treating a point M as a complex number

$$a_n = \frac{1}{L}\sum_{k=1}^{m}\frac{1}{l_k - l_{k-1}}\Big[(V_{k-1}l_k - V_k l_{k-1})\cdot\int_{l_{k-1}}^{l_k} \mathrm{e}^{-\mathrm{i}\frac{2\pi}{L}nl}\mathrm{d}l + (V_k - V_{k-1})\cdot\underbrace{\int_{l_{k-1}}^{l_k} l\cdot \mathrm{e}^{-\mathrm{i}\frac{2\pi}{L}nl}\mathrm{d}l}_{I}\Big]. \tag{2.18}$$

By computing

$$I = \int_{l_{k-1}}^{l_k} l\cdot\mathrm{e}^{-\mathrm{i}\frac{2\pi}{L}nl}\mathrm{d}l = -\frac{1}{\mathrm{i}\frac{2\pi}{L}n}\left(l_k\cdot\mathrm{e}^{-\mathrm{i}\frac{2\pi}{L}nl_k} - l_{k-1}\cdot\mathrm{e}^{-\mathrm{i}\frac{2\pi}{L}nl_{k-1}}\right)$$
$$+\frac{1}{\left(\frac{2\pi}{L}n\right)^2}\left(\mathrm{e}^{-\mathrm{i}\frac{2\pi}{L}nl_k} - \mathrm{e}^{-\mathrm{i}\frac{2\pi}{L}nl_{k-1}}\right)$$

and substituting it into (2.18) we shall achieve

$$a_n = \frac{1}{L}\sum_{k=1}^{m}\frac{1}{l_k - l_{k-1}}\cdot T_k, \tag{2.19}$$

where

$$T_k = -\frac{1}{\mathrm{i}\frac{2\pi}{L}n}(V_{k-1}l_k - V_k l_{k-1})\cdot\left(\mathrm{e}^{-\mathrm{i}\frac{2\pi}{L}nl_k} - \mathrm{e}^{-\mathrm{i}\frac{2\pi}{L}nl_{k-1}}\right)$$
$$-\frac{1}{\mathrm{i}\frac{2\pi}{L}n}(V_k - V_{k-1})\cdot\left(l_k\mathrm{e}^{-\mathrm{i}\frac{2\pi}{L}nl_k} - l_{k-1}\mathrm{e}^{-\mathrm{i}\frac{2\pi}{L}nl_{k-1}}\right)$$
$$+\frac{1}{\left(\frac{2\pi}{L}n\right)^2}(V_k - V_{k-1})\cdot\left(\mathrm{e}^{-\mathrm{i}\frac{2\pi}{L}nl_k} - \mathrm{e}^{-\mathrm{i}\frac{2\pi}{L}nl_{k-1}}\right),$$

namely

$$T_k = -\frac{1}{\mathrm{i}\frac{2\pi}{L}n}\left[V_k(l_k - l_{k-1})\cdot\mathrm{e}^{-\mathrm{i}\frac{2\pi}{L}nl_k} - V_{k-1}(l_k - l_{k-1})\cdot\mathrm{e}^{-\mathrm{i}\frac{2\pi}{L}nl_{k-1}}\right]$$
$$+\frac{1}{\left(\frac{2\pi}{L}n\right)^2}(V_k - V_{k-1})\cdot\left(\mathrm{e}^{-\mathrm{i}\frac{2\pi}{L}nl_k} - \mathrm{e}^{-\mathrm{i}\frac{2\pi}{L}nl_{k-1}}\right). \tag{2.20}$$

Therefore, on the basis of the relations (2.20) and (2.15) from which it results

$$l_k - l_{k-1} = |V_k - V_{k-1}|,\ (\forall)\ k = \overline{1, m},$$

the formula (2.19), which allows us to compute the Fourier descriptors will be:

$$a_n = -\frac{1}{2\pi i n} \sum_{k=1}^{m} \left(V_k \cdot e^{-i\frac{2\pi}{L} n l_k} - V_{k-1} \cdot e^{-i\frac{2\pi}{L} n l_{k-1}} \right)$$

$$+ \frac{L}{(2\pi n)^2} \sum_{k=1}^{m} \frac{V_k - V_{k-1}}{|V_k - V_{k-1}|} \cdot \left(e^{-i\frac{2\pi}{L} n l_k} - e^{-i\frac{2\pi}{L} n l_{k-1}} \right). \tag{2.21}$$

The main advantage of using the Fourier Descriptor method for object recognition consists in the fact that the Fourier descriptors are [45] invariant to translation, rotation, and scale.

2.3.5 *Fuzzy Neurons*

First, we shall define a Fuzzy Neuron (FN) in general and then the four models of fuzzy neurons, that are the components of the fuzzy neural network, described in this chapter.

A FN (illustrated in Fig. 2.6) has [8]:

- N weighted inputs $x_i,\ i = \overline{1, N}$
- N weights $w_i,\ i = \overline{1, N}$
- M outputs $y_j,\ j = \overline{1, M}$.

Each output can be associated with a membership value to a fuzzy concept, in the meaning that it expresses to what degree, the pattern with the inputs $\{x_1, x_2, \ldots, x_N\}$ belongs to a fuzzy set.

The equations that characterize a FN are [8]:

$$z = h(w_1 x_1, w_2 x_2, \ldots, w_N x_N), \tag{2.22}$$

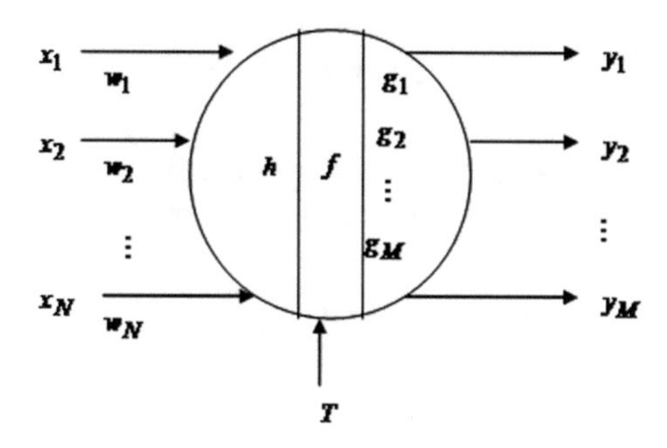

Fig. 2.6 A FN

$$s = f(z - T), \tag{2.23}$$

$$y_j = g_j(s),\ j = \overline{1, M}, \tag{2.24}$$

where

- z is the input net of the FN,
- h constitutes the aggregation function,
- s means the state of the FN,
- f is the activation function,
- T is the activating threshold,
- g_j, $j = \overline{1, M}$ are the M outputs of the FN, that represents the membership functions of the input pattern $\{x_1, x_2, \ldots, x_N\}$ to the M fuzzy sets.

Therefore, the FN can express and process the fuzzy information.

In general, the weights, the activating threshold, and the output functions that describe the interactions among the fuzzy neurons could be adjusted during the learning procedure. Hence, the fuzzy neurons are adaptive and a Fuzzy Neural Network (FNN) which has such neurons can learn from the environment.

The activation function and the activating threshold are the intrinsic features of a FN.

If different functions of f and h are used for different fuzzy neurons, then their properties will be different. Choosing different functions f and h we can achieve many types of fuzzy neurons.

The following four types of fuzzy neurons will be defined [8]:

(A) **Input-FN** is a FN, which is used in the input layer of a FNN and that has only one input x such that

$$z = x; \tag{2.25}$$

(B) **Maximum-FN (Max-FN)** is that FN, (represented in Fig. 2.7), which has a maximum function as a aggregation function, such that

$$z = \max_{i=1}^{N}(w_i x_i); \tag{2.26}$$

(C) **Minimum-FN (Min-FN)** is that FN (represented in Fig. 2.8), which has a minimum function as a aggregation function, such that

$$z = \min_{i=1}^{N}(w_i x_i); \tag{2.27}$$

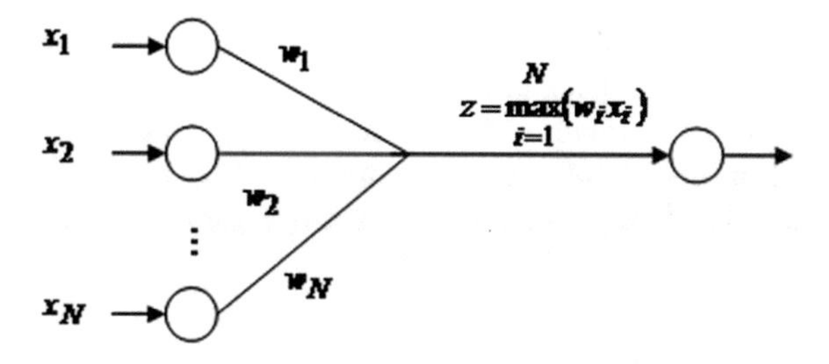

Fig. 2.7 A Max-FN

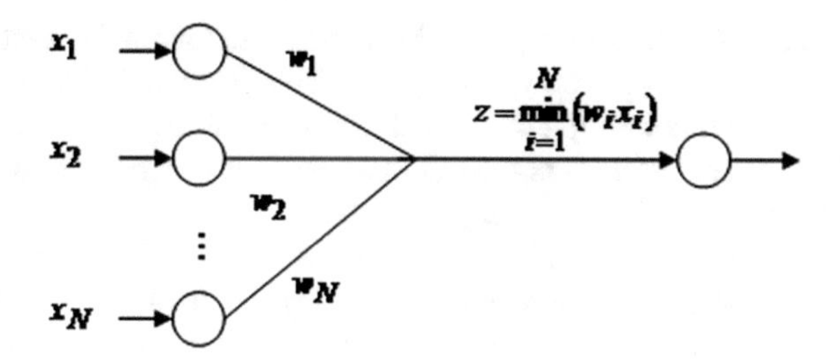

Fig. 2.8 A Min-FN

(D) **Competitive-FN (Comp-FN)** is a FN which has a variable threshold T and only one output, such that

$$y = g(s - T) = \begin{cases} 0 \text{ if } s < T, \\ 1 \text{ if } s \geq T. \end{cases} \tag{2.28}$$

$$T = t(c_1, \ldots, c_k), \tag{2.29}$$

where

- s is the state of the FN;
- t is a threshold function;
- $c_k,\ k = \overline{1, K}$ are the competitive variables of the FN.

2.4 Fuzzy Kwan–Cai Neural Network

In this study, we aim to provide an effective method for learning image similarity. To reach this aim we start from the Fuzzy Kwan–Cai Neural Network (FKCNN) and turn it into a supervised method for learning similarity. The new approach gives a better performance than its unsupervised counterparts [8] as in the case of the classical unsupervised FKCNN, a class is represented by a single-output neuron, while the supervised FKCNN has more output neurons that each designate a class. This concept is similar with the idea of replacing the binary decision about the membership (nonmembership) of a pattern to a class, with the introduction of a membership degree, between 0 and 1.

Using the following four types of fuzzy neurons [8]: **Input-FN**, **Maximum-FN (Max-FN)**, **Minimum-FN (Min-FN)**, **Competitive-FN (Comp-FN)** as basis we can develop our Fuzzy Kwan–Cai neural network [8] for learning similarity.

2.4.1 Architecture of FKCNN

The FKCNN is a feedforward fuzzy neural network with four layers each containing a specific type of neuron, as shown in Fig. 2.9. This is a parallel system designed both for image recognition [30] and for the Information Retrieval: it simultaneously processes all the pixels of an input image, namely it is a parallel system. Its each layer is built by a specific type of fuzzy neurons.

The first layer of the network is the input layer, which ensures the provision of patterns, that the network has to recognize. This layer contains Input-FNs and each neuron corresponds to a feature of the input pattern. The Input-FNs are arranged in a 2D structure and their number is equal to the number of features that belong to the input pattern. Assuming that each pattern has $N_1 \times N_2$ features, it results that the first layer has $N_1 \times N_2$ neurons. The algorithm of the (i, j)-th Input-FN in the first layer is

$$s_{ij}^1 = z_{ij}^1 = x_{ij}, \ (\forall)\ i = \overline{1, N_1},\ j = \overline{1, N_2}, \tag{2.30}$$

$$y_{ij}^1 = \frac{s_{ij}^1}{P_{vmax}}, \ (\forall)\ i = \overline{1, N_1},\ j = \overline{1, N_2}, \tag{2.31}$$

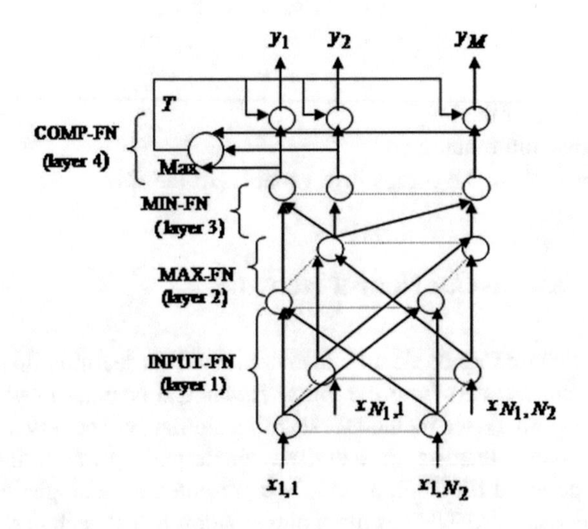

Fig. 2.9 Architecture of FKCNN

where x_{ij} is the (i, j)-th feature of an input pattern ($x_{ij} \geq 0$) and P_{vmax} is the maximum feature among all input patterns.

From Eq. (2.31) we can notice that the first layer receives image data and transforms the features corresponding to the input image into normalized values within the interval [0, 1].

The second layer is also organized as a 2D grid and consists of $N_1 \times N_2$ Max-FNs. The purpose of this layer is to fuzzy the input patterns through a weight function $w(m, n)$.

The state of the (p, q)-th Max-FN in this layer is given [8] by

$$s_{pq}^{[2]} = \max_{i=1}^{N_1} \left(\max_{j=1}^{N_2} \left(w(p-i, q-j) y_{ij}^{1} \right) \right), \ (\forall)\, p = \overline{1, N_1},\ q = \overline{1, N_2}, \qquad (2.32)$$

where $w(p-i, q-j)$ is the weight connecting the (i, j)-th Input-FN in the first layer to the (p, q)-th Max-FN of the second layer.

The weight function has [8] the following analytical expression:

$$w(m, n) = e^{-\beta^2 (m^2+n^2)}, \ (\forall)\, m = \overline{-(N_1-1), (N_1-1)},\ n = \overline{-(N_2-1), (N_2-1)}. \qquad (2.33)$$

Using the fuzzification weight function given by (2.33), "each Max-FN in the second layer behaves like a lens focusing on a specific feature of the input pattern."[9]

It can keep into account different neighbor features of the central one, too. The parameter β controls the number of features seen by a Max-FN and its value is determined by the learning algorithm.

Each FN in the second layer has M different outputs, one for each FN of the third layer. The outputs of the (p, q)-th Max-FN in the second layer are given [8] by the relation:

$$y_{pqm}^{[2]} = g_{pqm}(s_{pq}^{[2]}), \ (\forall)\, p = \overline{1, N_1},\ q = \overline{1, N_2},\ m = \overline{1, M}, \qquad (2.34)$$

where $y_{pqm}^{[2]}$ is the m-th output of the (p, q)-th Max-FN in the second layer, that is connected to the m-th Min-FN in the third layer.

The output function $g_{pqm}(s_{pq}^{[2]})$ will be determined through a learning algorithm. For simplicity, the fuzzy output functions of neurons in the second layer will be commonly chosen as isosceles triangles, having the height equal to 1 and the base length equal to α (see Fig. 2.10).

[9]Kwan, H. K. and Cai, Y., A Fuzzy Neural Network and its Application to Pattern Recognition, IEEE Trans. on Fuzzy Systems, 1997, 2(3), 185–193.

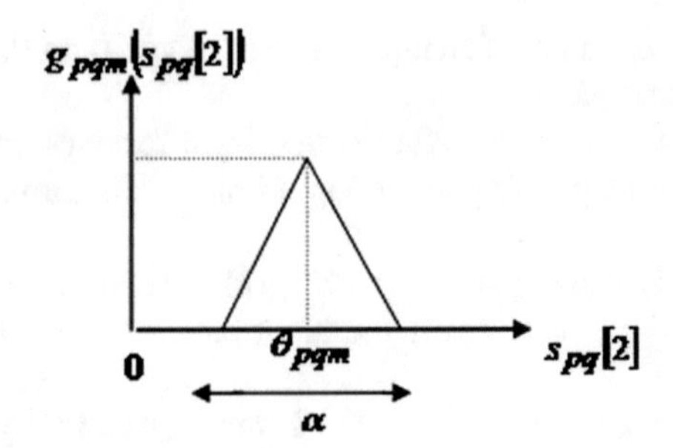

Fig. 2.10 The output function of a Max-FN

Then [8]:

$$y_{pqm}^{[2]} = g_{pqm}(s_{pq}^{[2]}) = \begin{cases} 1 - 2|s_{pq}^{[2]} - \theta_{pqm}|/\alpha \text{ if } \alpha/2 \geq |s_{pq}^{[2]} - \theta_{pqm}| \geq 0, \\ 0 \text{ otherwise,} \end{cases} \tag{2.35}$$

$(\forall)\ \alpha \geq 0,\ p = \overline{1, N_1},\ q = \overline{1, N_2},\ m = \overline{1, M}$, θ_{pqm} being the central point of the base of function $g_{pqm}(s_{pq}^{[2]})$.

The third layer contains Min-FNs; as each of them represents a learnt pattern, the number M of the Min-FNs in the third layer could be determined only after the learning procedure is finished. The output of the m-th Min-FN in the third layer is given [8] by the relation:

$$y_m^{[3]} = s_m^{[3]}) = \min_{p=1}^{N_1} \left(\min_{q=1}^{N_2} (y_{pqm}^{[2]}) \right),\ (\forall)\ m = \overline{1, M}, \tag{2.36}$$

where $s_m^{[3]}$ represents the state of the m-th Min-FN, in the third layer.

The fourth layer is the output layer and it contains M Comp-FNs, one for each of the M learnt patterns. It provides nonfuzzy (crisp) outputs. If an input image is most similar to the m-th learnt pattern, then the output of the m-th Comp-FN becomes 1, while the other outputs are equal to 0. The relations that characterize the Comp-FN in the fourth layer are [8]:

$$s_m^{[4]} = z_m^{[4]}) = y_m^{[3]},\ (\forall)\ m = \overline{1, M}, \tag{2.37}$$

$$y_m^{[4]} = g(s_m^{[4]} - T) = \begin{cases} 0, \text{ if } s_m^{[4]} < T \\ 1, \text{ if } s_m^{[4]} = T, \end{cases} \tag{2.38}$$

$(\forall)\ m = \overline{1, M}$. The activation threshold T of all the Comp-FNs in the fourth layer is

$$T = \max_{m=1}^{M}(y_m^{[3]}), \ (\forall)\ m = \overline{1, M}. \tag{2.39}$$

2.4.2 Training Algorithm of FKCNN

In the learning phase, the following parameters will be determined by the learning procedure:

- the parameters of the output functions of the Max-FNs in the second layer, namely θ_{pqm}, $(\forall)\ p = \overline{1, N_1},\ q = \overline{1, N_2},\ m = \overline{1, M}$,
- the number of neurons for the third and the fourth layers.

To do so, we define a parameter T_f, $0 \leq T_f \leq 1$ that represents the fault tolerance of the FKCNN. Let K be the total number of training patterns. For $k = \overline{1, K}$, the steps of the learning procedure are:

Step 1 Create $N_1 \times N_2$ Input-FNs in the first layer of the FKCNN and $N_1 \times N_2$ Max-FNs in the second layer. Choose a value for α, $\alpha \geq 0$ and a value for β.

Step 2 Set $M = 0$ and $k = 1$.

Step 3 Set $M = M + 1$. Create the M-th Min-FN in the third layer and the M-th Comp-FN in the fourth layer. Determine the central point corresponding to the M output functions for the (p, q)-th Max-FN in the second layer, denoted θ_{pqM}, using the formula:

$$\theta_{pqM} = s_{pqM}^{[2]} = \max_{i=1}^{N_1}\left(\max_{j=1}^{N_2}\left(w(p-i, q-j)x_{ijk}\right)\right), \ (\forall)\ p = \overline{1, N_1},\ q = \overline{1, N_2}, \tag{2.40}$$

$X_k = \{x_{ijk}\}$ being the k-th training pattern.

Step 4 Set $k = k+1$. If $k > K$ the learning algorithm finishes. Otherwise, introduce the k-th training pattern to network input and compute the outputs of the fourth layer of the FKCNN (with M Min-FNs in the third layer and M Comp-FNs in the fourth layer). Set

$$\sigma = 1 - \max_{j=1}^{M} y_{jk}^{[3]}, \tag{2.41}$$

where $y_{jk}^{[3]}$ is the output of the j-th Min-FN in the third layer, for the k-th training pattern X_k.

If $\sigma \leq T_f$ then go to *Step 4*. If $\sigma > T_f$ then go to *Step 3*.

In one training epoch, all patterns in the training set are sequentially given as input to the FKCNN.

The procedure of transforming the unsupervised fuzzy neural network into a supervised one is similar to that used in several applications of the Self-Organizing Map (Kohonen) [47, 48].

After the unsupervised learning algorithm, one calibrates [30] the FKCNN to learn similarity through partial supervision. For example, if we have a training set consisting of K images, belonging to P classes, one assumes that after training, the number of FNs of the third and fourth levels is M.

Then, after calibration, each of the M outputs corresponding to the Comp-FNs has a label, representing one of the P classes (generally $K \geq M \geq P$). The FNs of the fourth layer are then labeled as follows: neurons whose output is equal to 1 are labeled with the class corresponding to the corresponding input pattern.

Classification of an image with unknown class is achieved through the association of the input pattern with the class corresponding to that neuron in the fourth layer which has the output equal to 1.

2.4.3 Analysis of FKCNN

The first level of the FKCNN has to input, "the feature glows of the respective pattern and transmits these signals to the second layer, into normalized values, from the interval [0, 1]."[10] The aim of the second layer is to fuzzify the input pattern. Each Max-FN in the second layer is connected to all the Input-FNs of the first layer through the weight function $w(m, n)$ and the state of a fuzzy Max-FN is given by the maximum of the weighted inputs.

The second layer of the FKCNN performs the fuzzification of the lower level features belonging to the input pattern. The degree of fuzzification of the input pattern through the second layer depends on the parameter β. The smaller β is, the more Max-FN in the second layer are affected by a lower level feature of the input pattern. If β is too small, then FKCNN can not separate some distinct training patterns, while if the β is too large the network loses its ability to recognize some displaced or distorted patterns. The values of the parameter β should be chosen such that the FKCNN can separate all the distinct training patterns and moreover the network has to have an acceptable recognition rate.

Each Max-FN from the second layer has M different outputs; therefore M will define the number of the FNs in the third layer. The m-th output of the (p, q)-th Max-FN neuron in the second layer, i.e., $y_{pqm}^{[2]}$ "expresses to what extent the fuzzy concept about that zone the component values around the (p, q)-th component of the input pattern are similar to the component values around the (p, q)-th component of the m learnt pattern." (see footnote 10).

The output function $g_{pqm}(s_{pq}^{[2]})$ is a membership function of this fuzzy set and it contains some data with regard to the component values around the (p, q)-th component

[10] Kwan, H. K. and Cai, Y., A Fuzzy Neural Network and its Application to Pattern Recognition, IEEE Trans. on Fuzzy Systems, 1997, 2(3), 185–193.

of the m pattern, which it has already learnt. The training algorithm uses θ_{pqM} at *Step 4*, to store such data; hence FKCNN can remember all learnt forms.

The fuzzy neurons of the third layer allows to compute the similarities of the input pattern with all the learnt patterns. As in the third layer one uses Min-FNs, the similarity of the input pattern $X = \{x_{ijk}\}$ to the m-th learnt pattern can be computed using [8] the formula:

$$y_m^{[3]} = \begin{cases} \min_{p,q}\left(1 - 2|s_{pq}^{[2]} - \theta_{pqm}|/\alpha\right) \text{ if } \max_{p,q}\left(|s_{pq}^{[2]} - \theta_{pqm}|\right) \leq \alpha/2, \\ 0 \text{ otherwise}, \end{cases} \tag{2.42}$$

$(\forall)\ m = \overline{1, M}$ and $s_{pq}^{[2]}$ is the state of the (p, q)-th Max-FN in the second layer, when $X = \{x_{ijk}\}$ are given as network input. The relation (2.42) shows that α is a visibility parameter (scope parameter) and its value affects the computation of the similarities. When the input pattern X is one of the previously learnt patterns, then one of the similarities $y_m^{[3]}$, $m = \overline{1, M}$ will be equal to 1. In the case when the input pattern X is not any of the learnt patterns then all the M similarities will be less than 1.

The output layer of the FKCNN is used to produce the defuzzification and to give some nonfuzzy outputs. The maximum similarity will be chosen as an activation threshold of all the Comp-FNs in the fourth layer. If $y_m^{[3]}$ has the maximum value among all the outputs of the FNs in the third layer, then the output of m-th Comp-FN in the fourth layer is equal to 1 and the outputs of the other Comp-FNs in this layer will be equal to 0.

The learning algorithm is developed in four stages: input data (layer 1), fuzzification (layer 2), fuzzy deduction (layer 3), and defuzzification (layer 4).

When a previously learned pattern is given as input, the network will treat the respective pattern as a known pattern, without relearning it. The way of treating an input pattern (either distinct from all patterns or a learned one) is determined both by the similarities of the FKCNN, computed to all the learnt patterns and the parameters α, β and T_f: α and β affect the computing of the similarities and T_f is the fault tolerance of the FKCNN. If one of the computed similarities to the input pattern is greater than or equal to $1 - T_f$, then this pattern will be treated as a learnt pattern; otherwise the pattern is treated as a new form.

2.5 Experimental Evaluation

2.5.1 Data Sets

We shall use the images from the PASCAL data set to evaluate our methods. The PASCAL Visual Object Classes (VOC) challenge is [9] a benchmark in visual object category recognition and detection, providing the vision and machine learning communities with a standard data set of images and annotation. In our work we use 10102 images half of them (chosen randomly) for training and the other half for testing. The VOC data sets contain significant variability in terms of object size, orientation, pose, illumination, position, and occlusion. It consists of 20 object classes: aeroplane, bicycle, bird, boat, bottle, bus, car, cat chair, cow, dog, horse, motorbike, person, sheep, sofa, table, potted plant, train, and tv/monitor. The ColorDescriptor engine [49] was used to extract the image descriptors from all the images. To evaluate our methods we use images from the PASCAL data set. The PASCAL Visual Object Classes (VOC) challenge is [9] a benchmark in visual object category recognition and detection, providing the vision and machine learning communities with a standard data set of images and annotation. In our work, we use 10102 images half of them for training and the other half for testing. Figure 2.11 shows 50 images from the VOC data base.

In our work [1], we have used 64 descriptors for each image from the training and test set. Hence, the number of the neurons of the first layer corresponding to FKCNN is 8×8 and 64 in the case of SOKM.

2.5.2 Evaluation Criteria

We aim to compare the performance of the three methods we have defined k-NN, SOKM, and FKCNN in terms of their ability to learn similarity. Following (2.2) we need [1] the following stages:

Stage 1 (compute the similarities among the patterns). For each test image T_k, $(\forall)\ k = \overline{1,\ 5051}$, classified in the class C_i', $(\forall)\ i = \overline{1,\ M}$ we compute its similarities S_{kp} $(\forall)\ k, p = \overline{1,\ 5051}$ regarding the training images $I_1, \ldots, I_{5051}$, that belong to the classes $C_1, \ldots, C_M$.

Stage 2 (find the similarities among the classes). Determine with (2.2) the similarity R_{ij} between the classes C_i' and C_j (corresponding to the test image T_k and respectively associated to a training image I_p $(\forall)\ p = \overline{1,\ 5051}$).

Stage 3 (evaluation criteria). Check if for $S_{k,p_j} \leq S_{k,p_l}$ we have $R_{ij} \leq R_{il}$ or when $S_{k,p_j} > S_{k,p_l}$ we have $R_{ij} > R_{il}$, i being the class where the test image T_k belongs and j respectively l being the classes corresponding to the two training images I_{p_j} and I_{p_l}.

To determine how well the methods perform we have to [1]:

- apply the procedure for all $N = 5051 \cdot (5051 - 1)$ pairs of images;
- find the percentage for which the similarities were correctly computed, using the formula:

$$p\,[\%] = \frac{n}{N} \cdot 100,$$

n being the number of correctly evaluated images in the view of their similarities.

Fig. 2.11 50 images from the VOC data base

2.5.3 Experimental Results

For our experiments we set $\alpha = 1,\ \beta = 1$ as the parameters of the FKCNN. With different values of the threshold T_f we obtain the results in Table 2.2 and Fig. 2.12:

Figure 2.12 shows that the best value of M (corresponding to the number of neurons from the third and fourth layers of FKCNN) is 22 which is close to the number of classes in the data set which is 20.

In the case of using SOKM (trained during some epochs) we evaluate the performance of the algorithm for computing similarities in the following way:

(1) find the coordinates of the winning neuron;
(2) compute the similarity between a test image and a training image based on the formula (2.6), where W_j is the weight vector of that neuron associated to the respective training image;

Table 2.2 The overall performance of FKCNN: the values of p [%] obtained using the FKCNN, for different values of T_f, when $\alpha = 1,\ \beta = 1$

T_f	M	p [%]
0.41	16	99.8
0.396	19	99.7
0.39	20	99.8
0.38	21	99.8
0.37	22	99.9
0.36	23	99.8
0.35	24	99.8
0.3212	30	99.7

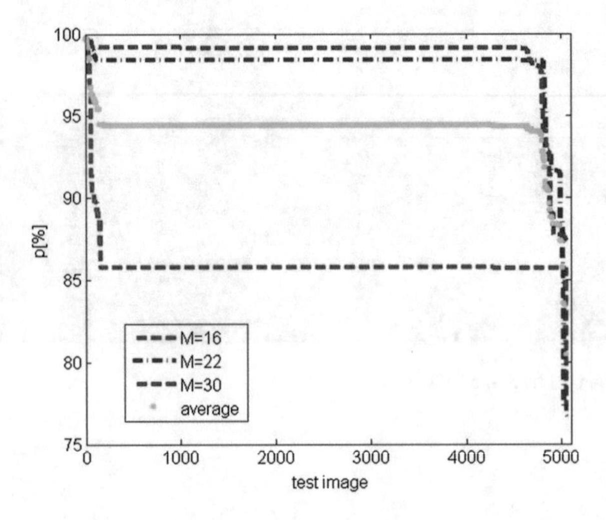

Fig. 2.12 The percentages for which the similarities were correctly computed using FKCNN

(3) check if we achieve a smaller similarity for a training image that is closer to that test image.

We want to determine the robustness with respect to choosing the right parameters.

Table 2.3 and respectively Table 2.4, Figs. 2.13 and 2.14 illustrate the percentages for which the similarities were correctly computed using SOKM, as function of number of epochs used in training epochs and respectively by changing the weights.

Although, sometimes we can achieve good results using SOKM, it has an unstable behavior as Figs. 2.13 and 2.14 show. We note that the instability of SOKM is determined by the following two aspects: by training the SOKM in five epochs, the value of p [%] changes, especially in the cases when the weights increase; in these cases, p [%] will have smaller values than the values achieved when the weights decrease; in the situations where the SOKM is trained with a different number of epochs, the values of p [%] will fluctuate from one epoch to another.

Figure 2.15 and Table 2.5 indicate that for our task FNNKC and SOKM perform better than the nonneural method k-NN, with 22 neighbors.

We observe that:

- the nonneural k-NN method computes more wrong similarities than the neural methods;
- using the FKCNN we achieved the biggest number of correct similarities (i.e., 5038) between a test image and the training images compared to the other two methods;
- FKCNN is also most performant as the biggest sum of the number of correct similarities is achieved by it (namely, for each test image we achieved a big number

Table 2.3 The percentages for which the similarities were correctly computed using SOKM, after some training epochs

Nrep	p [%]
1	77
2	92.2
5	98.4
10	75.3
15	75.4
25	70.5

Table 2.4 The percentages for which the similarities were correctly computed using SOKM, by changing the weights

Interval where the weight belongs	p [%]
[−0.002, 0.002]	97.9
[−0.2, 0.2]	72.2
[−0.05, 0.05]	73
[−0.006, 0.006]	94.9
[−0.09,0.09]	73.6

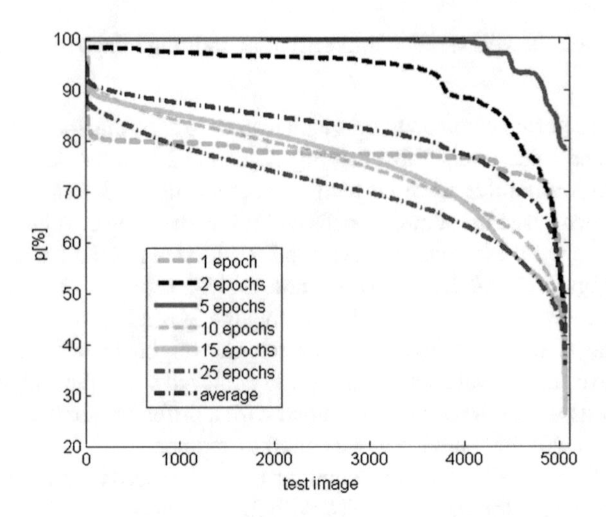

Fig. 2.13 The percentages for which the similarities were correctly computed using SOKM, after some training epochs

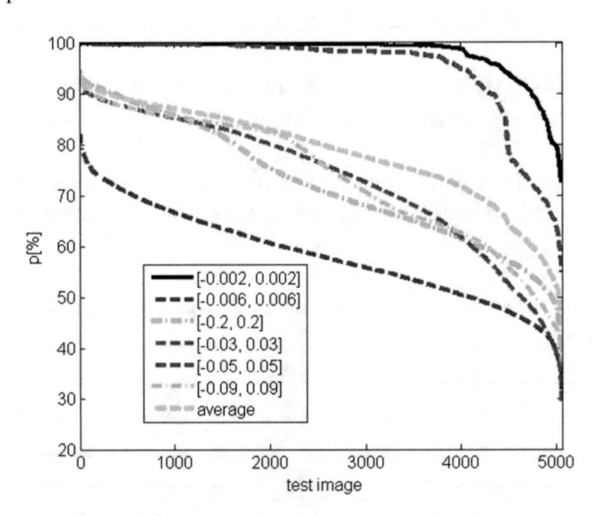

Fig. 2.14 The percentages for which the similarities were correctly computed using SOKM, by changing the weights

of correct similarities with FKCNN, while using SOKM we can not have almost the same number of correct similarities for each image);

- the graph corresponding to the k-NN method is an almost constant function, while the functions of the neural methods have a descending behavior.

In Fig. 2.16 we can notice that the FKCNN is always better than SOKM as we have obtained the maximum value of $p = 99.86\,\%$ using FKCNN (see Table 2.2). The SOKM has better performance for the easy elements, the ones that are in the

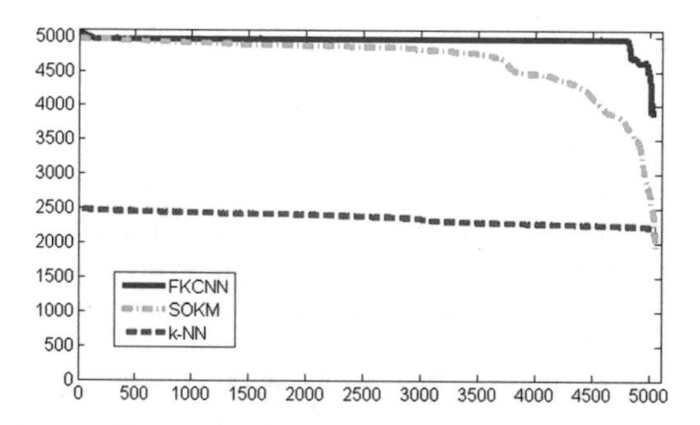

Fig. 2.15 The overall performance of the three proposed methods for computing similarities

Table 2.5 The percentage of those test images for which works well our procedure (for $M = 20$)

Used model	p [%]
FKCNN	99.75
k-NN	72.2
Euclidean distance (Cosine similarity)	50.12 (50.14)
SOKM	98.80

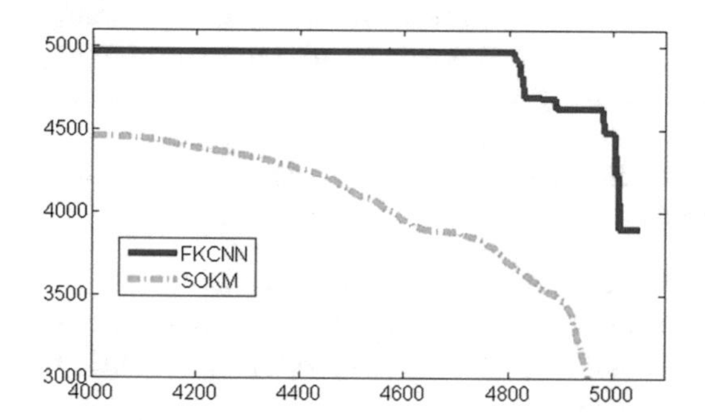

Fig. 2.16 Close-up of the overall performance of Fig. 2.15 of the neural methods for computing similarities

beginning of the list. FCKNN does a better job at the tail where the more difficult and likely more important elements in terms of tasks like clustering/classification lie.

The results indicate that the neural methods FKCNN and SOKM are performing better for our task than k-NN (see Fig. 2.15). SOKM sometimes gives good results, but this depends highly on the right parameter settings. Small variations induced large drops in performance. The overall performance of FKCNN is better (see Fig. 2.16). The main advantage of FKCNN consists in the fact that we can obtain good results that are robust to changes in the parameter settings.

2.6 Face Recognition

Face Recognition (FR) has been studied for many years and has practical and/or potential applications in areas such as security systems, criminal identification, video telephony, medicine, and so on. In comparison to other identification techniques, FR has the advantage of being nonintrusive and requiring very little cooperation; it has become a hot topic recently as better hardware and better software have become available.

On the other side, the hybrid systems of *fuzzy logic* and *neural networks* (often referred as *fuzzy neural networks*) represent exciting models of computational intelligence with direct applications in pattern recognition, approximation, and control.

Face Recognition represents a segment of a larger technological areas called Biometric Technology (BT). The BT has modern practical applications in the following fields: Medicine, Justice and Police, Finance and Banks, Border control, Voice and Visual communications, Access control for sensitive areas, as indicated [50] in Fig. 2.17.

BT is widely used for the applications that include: face recognition, voice recognition, signature recognition, hand geometry, iris, Automated Fingerprint Identification System (AFIS), and non-AFIS.

As an industry, the biometrics is in its early stages; according to a study of the *Grand View Research* (a market research and consulting company),[11] the revenues for this industry have a significant growth, such that it is expected to reach USD 24.59 billion by 2020 (see Fig. 2.18).

FR is an automated approach to identify humans through the unique features of their faces. It needs

(A) a camera to take a digital image of a subject's face;
(B) some algorithms to extract the facial features from the respective image, resulting a template. Various approaches have been proposed [51] to select the facial features from an image:

  (a) *Geometry based Technique*-the selection is based on the size and the relative position of important components of images.

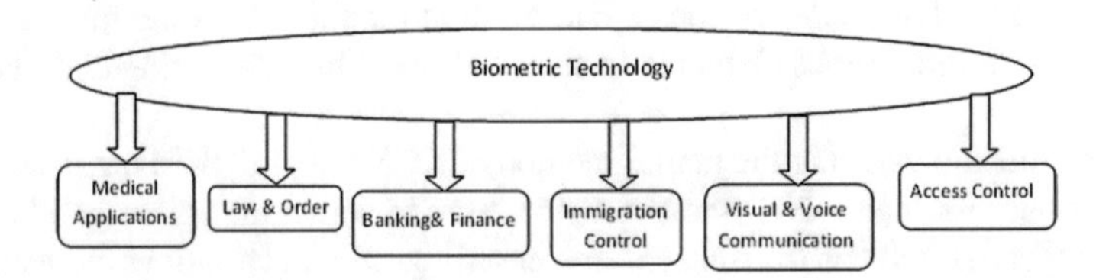

Fig. 2.17 Main modern BT applications

[11] Grand View Research, 2014, http://www.grandviewresearch.com/industry-analysis/biometrics-industry.

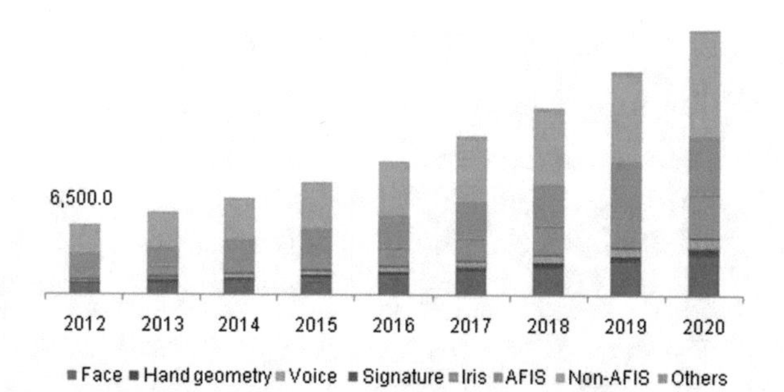

Fig. 2.18 Global biometrics technology market by application (USD Million), 2012–2020 (see footnote 11)

(b) *Appearance based approach* has the main purpose to keep the important information of image and reject the redundant information, like examples being the Principal Component Analysis (PCA), Discrete Cosine Transformation (DCT), Independent Component Analysis (ICA), Linear Discriminant Analysis (LDA).
(c) *Template Based Techniques* uses an appropriate energy function to extract the facial features based on the previously designed templates; the minimum energy is achieved for the best match of the template in facial image.
(d) *Color based approach*—the skin color is used to isolate the face area from the nonface area in an image.

(C) the comparison of the template formed at step (B) to the templates in the considered database.

The performance of the face recognition systems is affected by a lot of factors such as:

- variations in lighting conditions;
- camera angles;
- changes in position and expression of the face.

A very important stage of the FR process is called *eigenfaces* and it consists in the determination of the best facial features, which discriminates the features of a subject from those of another face.

We have experimented [30] our model using the *ORL Database of Faces*, provided by the AT&T Laboratories from Cambridge University.

The database has 400 images, corresponding to 40 subjects (namely, 10 images for each of the 40 classes). We have divided the whole gallery into a training lot (200 pictures) and a test lot (200 pictures). Each image has the size of 92×112 pixels with 256 levels of gray. For the same subject (class), the images have been taken at different hours, lighting conditions, and facial expressions, with or without glasses. For each class, one chooses five images for training and five images for test.

Figure 2.19 illustrates 100 images from the ORL face database.

Fig. 2.19 100 images from the ORL face database

2.6.1 *Applying the Fuzzy Kwan–Cai Neural Network for Face Recognition*

We have performed the software implementation of the FKCNN to be experimented for the face recognition task, both in the case of 100 images (10 classes) [30, 52], and 400 images (40 classes) [52], too. A half images from the used database constitutes the training lot of the FKCNN and the other half, its test lot.

The images belonging to the training and respectively from the test lot (being different from those that are in training lot) have been applied alternatively and in only one learning epoch.

After the training algorithm of the FKCNN, it is necessary to preserve its parameters and go to the network calibration (partial supervision). This stage also means a testing on the training lot (normally, we have to obtain here a recognition score of 100 %).

The output neurons that yield the maximum value 1 constitutes the label corresponding to the training input image.

We need the following algorithm to classify a test image:

(1) find the label of that neuron of the fourth level of the FKCNN, whose output is equal to 1;
(2) check if the label of the neuron coincides with the label of the respective input image to see if the recognition is correctly; otherwise it is erroneously.

The score of the correct recognition can be computed using the formula:

$$R\,[\%] = \frac{\textit{the number of the correct recognitions}}{\textit{the total number of the test images}} \times 100. \tag{2.43}$$

Remark 2.2 The same procedure and formula is also applied for the training lot, in the calibration stage to validate the FKCNN learning.

The following parameters are chosen in order to guarantee a good behavior of the learning algorithm:

- the parameter α characterizing the nonlinear output functions of the MAX-FN's in the second layer;
- the parameter β, characterizing the fuzzification function;
- T_f-the fault tolerance of the FKCNN.

We shall further evaluate the effects of the incoming parameters α, β, and T_f on the recognition rate and on the number M of output neurons obtained during the learning procedure.

Some experimental results are given in Table 2.6.

From Table 2.6 we can evaluate that the very good recognition score of 94 % is obtained by the FKCNN on the test lot of 50 images, for the following two optimum combinations of parameters:

(1) $\alpha = 1.5$, $\beta = 0.014$, $T_f = 0.052$;
(2) $\alpha = 2$, $\beta = 0.014$, $T_f = 0.039$.

Table 2.6 The recognition rate R [%] of the FKCNN on the test lot (50 images of 10 classes) as a function of the parameters α, β, and T_f

α	β	T_f	R [%]
1.5	0.015	0.076	90
	0.0145	0.061	92
	0.014	**0.052**	**94**
	0.0135	0.051	92
	0.013	0.048	84
2	0.015	0.057	90
	0.0145	0.045	92
	0.014	**0.039**	**94**
	0.0135	0.038	92
	0.013	0.036	84

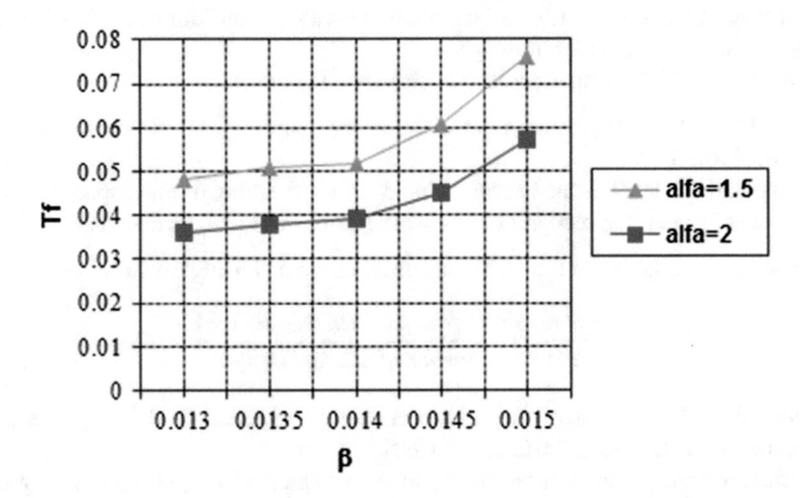

Fig. 2.20 The fault tolerance T_f as a function of the parameter β, in order to obtain good recognition rates

The fault tolerance T_f has been chosen in order to obtain a number of MIN-FN's on the third layer equal to the number of images belonging to the training lot.

The FKCNN is very sensitive [30] to the selection of the parameter β. For example, if $\alpha = 2$ then a modification of the parameter β with only 0.001 (from 0.014 to 0.013) produces the decreasing of the recognition rate with 10 % (from 94 to 84 %).

Conversely, the parameter $\alpha = 2$ has a reduced influence on the recognition score.

From Fig. 2.20 we can notice that to obtain good recognition rates (for example $R \geq 84$ %, according to the results given in Table 2.6), when the parameter β increases, one has to increase the fault tolerance T_f, too.

The matrix C shows the confusion matrix for the best recognition score ($R = 94\,\%$, $\alpha = 2$, $\beta = 0.014$, $T_f = 0.039$).

$$
\begin{array}{c}
\\ \\ \\ \\ \\
\mathbf{C} =
\end{array}
\begin{array}{c}
\textit{Assigned class} \\ 1 \\ 2 \\ 3 \\ 4 \\ 5 \\ 6 \\ 7 \\ 8 \\ 9 \\ 10
\end{array}
\begin{array}{c}
\textit{Real class} \quad 1 \;\; 2 \;\; 3 \;\; 4 \;\; 5 \;\; 6 \;\; 7 \;\; 8 \;\; 9 \;\; 10 \\
\begin{pmatrix}
80 & 0 & 0 & 0 & 0 & 0 & 20 & 0 & 0 & 0 \\
0 & 100 & 0 & 0 & 0 & 0 & 0 & 0 & 0 & 0 \\
0 & 0 & 100 & 0 & 0 & 0 & 0 & 0 & 0 & 0 \\
0 & 0 & 20 & 80 & 0 & 0 & 0 & 0 & 0 & 0 \\
0 & 0 & 0 & 0 & 80 & 20 & 0 & 0 & 0 & 0 \\
0 & 0 & 0 & 0 & 0 & 100 & 0 & 0 & 0 & 0 \\
0 & 0 & 0 & 0 & 0 & 0 & 100 & 0 & 0 & 0 \\
0 & 0 & 0 & 0 & 0 & 0 & 0 & 100 & 0 & 0 \\
0 & 0 & 0 & 0 & 0 & 0 & 0 & 0 & 100 & 0 \\
0 & 0 & 0 & 0 & 0 & 0 & 0 & 0 & 0 & 100
\end{pmatrix}
\end{array}
$$

The matrix C illustrates that in the case when $\alpha = 2$, $\beta = 0.014$, $T_f = 0.039$, the FKCNN correctly recognizes 47 images of the 50 from the test lot.

In Fig. 2.21 are presented the labels that FKCNN assigned to the images of the test lot.

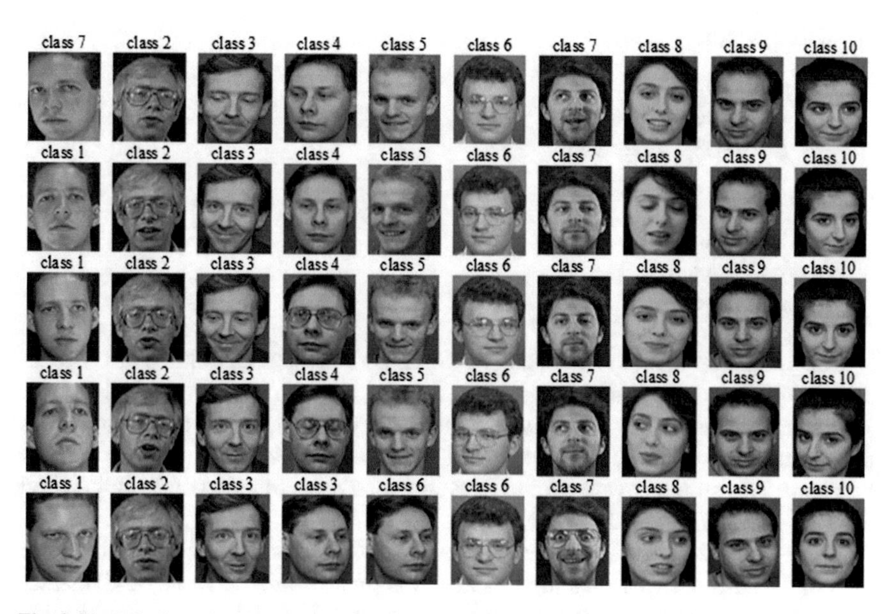

Fig. 2.21 Labeling of test lot for $\alpha = 2$, $\beta = 0.045$, $T_f = 0.114$

Table 2.7 The recognition rate R [%] of the FKCNN on the test lot (200 images of 40 classes) as a function of the parameters α, β, and T_f

α	β	T_f	R [%]
1.5	0.05	0.167	80.5
	0.045	**0.153**	**82.5**
	0.004	**0.133**	**82.5**
	0.035	**0.114**	**82.5**
	0.03	0.096	79.5
2	0.05	0.125	80.5
	0.045	**0.114**	**82.5**
	0.04	**0.1**	**82.5**
	0.035	**0.085**	**82.5**
	0.03	0.072	79.5

Table 2.7 proves that we have also achieved [52] with the FKCNN, a recognition rate of 100 % over the training lot for a database of 40 classes, as in the hypothesis of using 10 classes (see Table 2.6).

From Table 2.7 we can notice that a good recognition score of 82.5 % can be achieved by the FKCNN on the test lot of 200 images, by choosing the following parameters:

(1) $\alpha = 1.5$ with

(a) $\beta = 0.045$, $T_f = 0.153$;
(b) $\beta = 0.04$, $T_f = 0.133$;
(c) $\beta = 0.035$, $T_f = 0.114$;

(2) $\alpha = 2$ with

(a) $\beta = 0.045$, $T_f = 0.114$;
(b) $\beta = 0.04$, $T_f = 0.1$;
(c) $\beta = 0.035$, $T_f = 0.085$;

The confusion matrix D depicted in Fig. 2.22 corresponds to the case: $\alpha = 2$, $\beta = 0.045$, $T_f = 0.114$ from Table 2.7.

Figure 2.23 shows the way how are chosen the values of the fault tolerance T_f as a function of β, for $\alpha = 1.5$ and $\alpha = 2$, in the case when $M = 40$.

Figure 2.24 shows how to choose the values of T_f as a function of β, for $\alpha = 1.5$ and $\alpha = 2$ ($M = 10$ and $M = 40$).

From Fig. 2.24 we can deduce that the FKCNN is less sensitive to the choice of β in the case of experimenting this neural approach for a database of 40 classes (consisting in 400 images) than when we applied the FKCNN for the classification of 10 classes of images.

Figure 2.25 presents the recognition rates over the test lot, being functions of the β values, for $\alpha = 1.5$ and $\alpha = 2$ ($M = 10$ and $M = 40$).

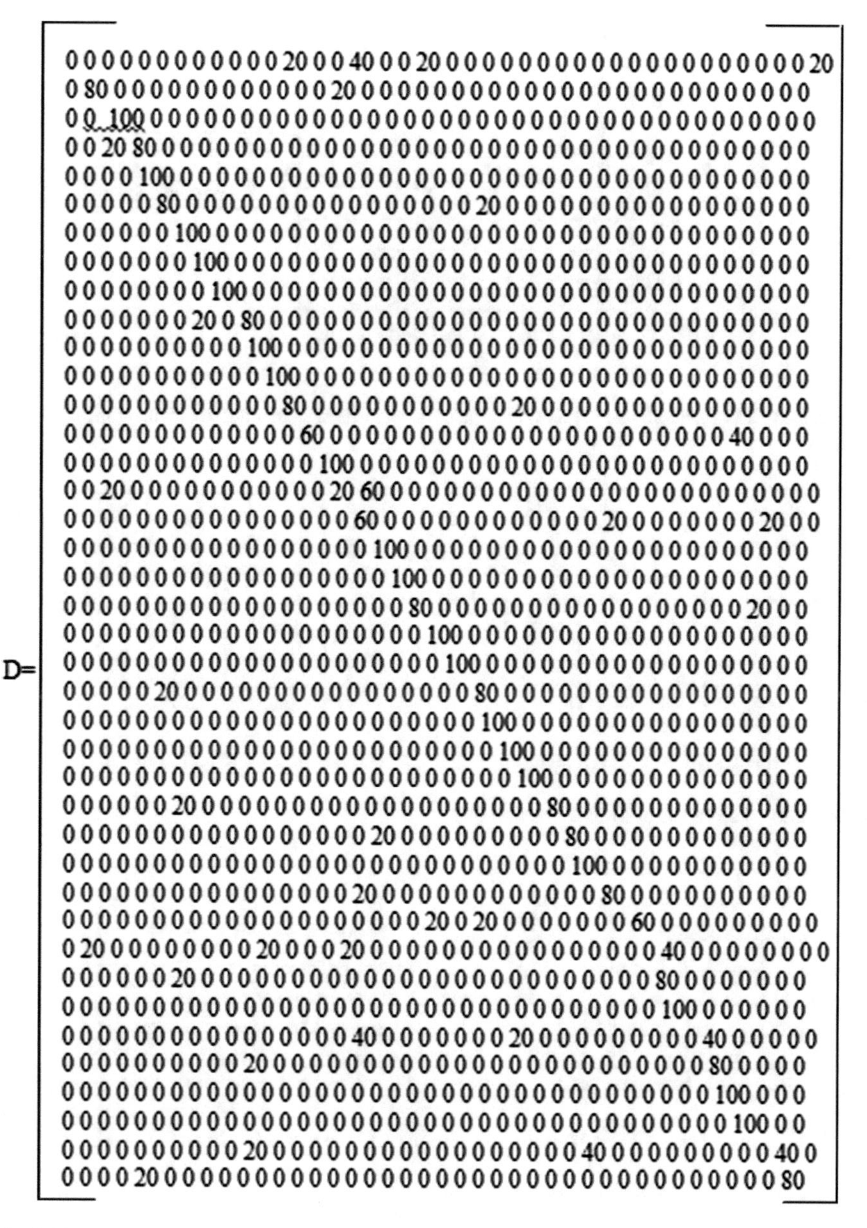

D=

0 0 0 0 0 0 0 0 0 0 0 0 20 0 0 40 0 0 20 0 0 0 0 0 0 0 0 0 0 0 0 0 0 0 0 0 0 0 0 20
0 80 0 0 0 0 0 0 0 0 0 0 0 0 20 0 0 0 0 0 0 0 0 0 0 0 0 0 0 0 0 0 0 0 0 0 0 0 0 0
0 0 100 0 0 0 0 0 0 0 0 0 0 0 0 0 0 0 0 0 0 0 0 0 0 0 0 0 0 0 0 0 0 0 0 0 0 0 0 0
0 0 20 80 0 0 0 0 0 0 0 0 0 0 0 0 0 0 0 0 0 0 0 0 0 0 0 0 0 0 0 0 0 0 0 0 0 0 0 0
0 0 0 0 100 0 0 0 0 0 0 0 0 0 0 0 0 0 0 0 0 0 0 0 0 0 0 0 0 0 0 0 0 0 0 0 0 0 0 0
0 0 0 0 0 80 0 0 0 0 0 0 0 0 0 0 0 0 0 0 0 0 20 0 0 0 0 0 0 0 0 0 0 0 0 0 0 0 0 0
0 0 0 0 0 0 100 0 0 0 0 0 0 0 0 0 0 0 0 0 0 0 0 0 0 0 0 0 0 0 0 0 0 0 0 0 0 0 0 0
0 0 0 0 0 0 0 100 0 0 0 0 0 0 0 0 0 0 0 0 0 0 0 0 0 0 0 0 0 0 0 0 0 0 0 0 0 0 0 0
0 0 0 0 0 0 0 0 100 0 0 0 0 0 0 0 0 0 0 0 0 0 0 0 0 0 0 0 0 0 0 0 0 0 0 0 0 0 0 0
0 0 0 0 0 0 0 20 0 80 0 0 0 0 0 0 0 0 0 0 0 0 0 0 0 0 0 0 0 0 0 0 0 0 0 0 0 0 0 0
0 0 0 0 0 0 0 0 0 0 100 0 0 0 0 0 0 0 0 0 0 0 0 0 0 0 0 0 0 0 0 0 0 0 0 0 0 0 0 0
0 0 0 0 0 0 0 0 0 0 0 100 0 0 0 0 0 0 0 0 0 0 0 0 0 0 0 0 0 0 0 0 0 0 0 0 0 0 0 0
0 0 0 0 0 0 0 0 0 0 0 0 80 0 0 0 0 0 0 0 0 0 0 0 20 0 0 0 0 0 0 0 0 0 0 0 0 0 0 0
0 0 0 0 0 0 0 0 0 0 0 0 0 60 0 0 0 0 0 0 0 0 0 0 0 0 0 0 0 0 0 0 0 0 0 0 40 0 0 0
0 0 0 0 0 0 0 0 0 0 0 0 0 0 100 0 0 0 0 0 0 0 0 0 0 0 0 0 0 0 0 0 0 0 0 0 0 0 0 0
0 0 20 0 0 0 0 0 0 0 0 0 0 0 20 60 0 0 0 0 0 0 0 0 0 0 0 0 0 0 0 0 0 0 0 0 0 0 0 0
0 0 0 0 0 0 0 0 0 0 0 0 0 0 0 0 60 0 0 0 0 0 0 0 0 0 0 0 0 20 0 0 0 0 0 0 0 20 0 0
0 0 0 0 0 0 0 0 0 0 0 0 0 0 0 0 0 100 0 0 0 0 0 0 0 0 0 0 0 0 0 0 0 0 0 0 0 0 0 0
0 0 0 0 0 0 0 0 0 0 0 0 0 0 0 0 0 0 100 0 0 0 0 0 0 0 0 0 0 0 0 0 0 0 0 0 0 0 0 0
0 0 0 0 0 0 0 0 0 0 0 0 0 0 0 0 0 0 0 80 0 0 0 0 0 0 0 0 0 0 0 0 0 0 0 0 0 20 0 0
0 0 0 0 0 0 0 0 0 0 0 0 0 0 0 0 0 0 0 0 100 0 0 0 0 0 0 0 0 0 0 0 0 0 0 0 0 0 0 0
0 0 0 0 0 0 0 0 0 0 0 0 0 0 0 0 0 0 0 0 0 100 0 0 0 0 0 0 0 0 0 0 0 0 0 0 0 0 0 0
0 0 0 0 0 20 0 0 0 0 0 0 0 0 0 0 0 0 0 0 0 0 80 0 0 0 0 0 0 0 0 0 0 0 0 0 0 0 0 0
0 0 0 0 0 0 0 0 0 0 0 0 0 0 0 0 0 0 0 0 0 0 0 100 0 0 0 0 0 0 0 0 0 0 0 0 0 0 0 0
0 0 0 0 0 0 0 0 0 0 0 0 0 0 0 0 0 0 0 0 0 0 0 0 100 0 0 0 0 0 0 0 0 0 0 0 0 0 0 0
0 0 0 0 0 0 0 0 0 0 0 0 0 0 0 0 0 0 0 0 0 0 0 0 0 100 0 0 0 0 0 0 0 0 0 0 0 0 0 0
0 0 0 0 0 0 20 0 0 0 0 0 0 0 0 0 0 0 0 0 0 0 0 0 0 0 80 0 0 0 0 0 0 0 0 0 0 0 0 0
0 0 0 0 0 0 0 0 0 0 0 0 0 0 0 0 0 20 0 0 0 0 0 0 0 0 0 80 0 0 0 0 0 0 0 0 0 0 0 0
0 0 0 0 0 0 0 0 0 0 0 0 0 0 0 0 0 0 0 0 0 0 0 0 0 0 0 0 100 0 0 0 0 0 0 0 0 0 0 0
0 0 0 0 0 0 0 0 0 0 0 0 0 0 0 0 20 0 0 0 0 0 0 0 0 0 0 0 0 80 0 0 0 0 0 0 0 0 0 0
0 0 0 0 0 0 0 0 0 0 0 0 0 0 0 0 0 0 0 0 20 0 20 0 0 0 0 0 0 0 60 0 0 0 0 0 0 0 0 0
0 20 0 0 0 0 0 0 0 0 20 0 0 0 20 0 0 0 0 0 0 0 0 0 0 0 0 0 0 0 0 40 0 0 0 0 0 0 0 0
0 0 0 0 0 0 20 0 0 0 0 0 0 0 0 0 0 0 0 0 0 0 0 0 0 0 0 0 0 0 0 0 80 0 0 0 0 0 0 0
0 0 0 0 0 0 0 0 0 0 0 0 0 0 0 0 0 0 0 0 0 0 0 0 0 0 0 0 0 0 0 0 0 100 0 0 0 0 0 0
0 0 0 0 0 0 0 0 0 0 0 0 0 0 0 0 40 0 0 0 0 0 0 0 20 0 0 0 0 0 0 0 0 0 40 0 0 0 0 0
0 0 0 0 0 0 0 0 0 0 20 0 0 0 0 0 0 0 0 0 0 0 0 0 0 0 0 0 0 0 0 0 0 0 0 80 0 0 0 0
0 0 0 0 0 0 0 0 0 0 0 0 0 0 0 0 0 0 0 0 0 0 0 0 0 0 0 0 0 0 0 0 0 0 0 0 100 0 0 0
0 0 0 0 0 0 0 0 0 0 0 0 0 0 0 0 0 0 0 0 0 0 0 0 0 0 0 0 0 0 0 0 0 0 0 0 0 100 0 0
0 0 0 0 0 0 0 0 0 0 20 0 0 0 0 0 0 0 0 0 0 0 0 0 0 0 0 0 40 0 0 0 0 0 0 0 0 0 40 0
0 0 0 0 20 0 0 0 0 0 0 0 0 0 0 0 0 0 0 0 0 0 0 0 0 0 0 0 0 0 0 0 0 0 0 0 0 0 0 80

Fig. 2.22 The confusion matrix corresponding to the parameters $\alpha = 2$, $\beta = 0.045$, $T_f = 0.114$

From Fig. 2.25 we can deduce that the FKCNN is very sensitive to the choice of parameter β, in the sense that a very small change corresponding to a value of β will contribute to the attainment of an other recognition rate over the test lot.

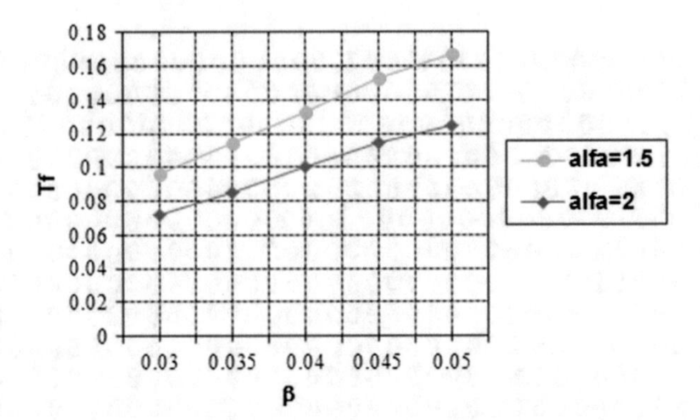

Fig. 2.23 The fault tolerance T_f as a function of the parameter β, in order to obtain good recognition rates

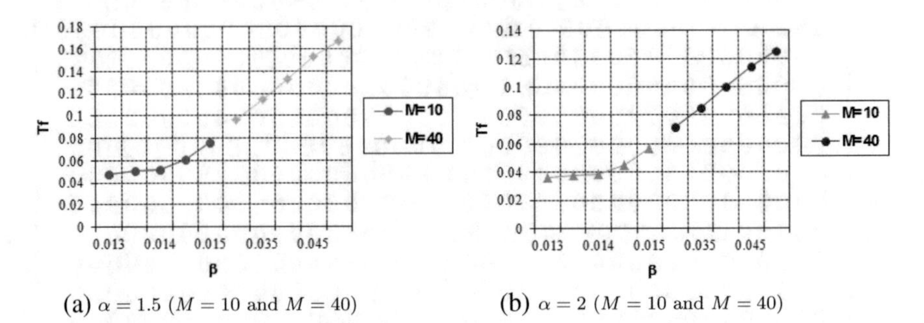

(a) $\alpha = 1.5$ ($M = 10$ and $M = 40$)

(b) $\alpha = 2$ ($M = 10$ and $M = 40$)

Fig. 2.24 The values of T_f as a function of β

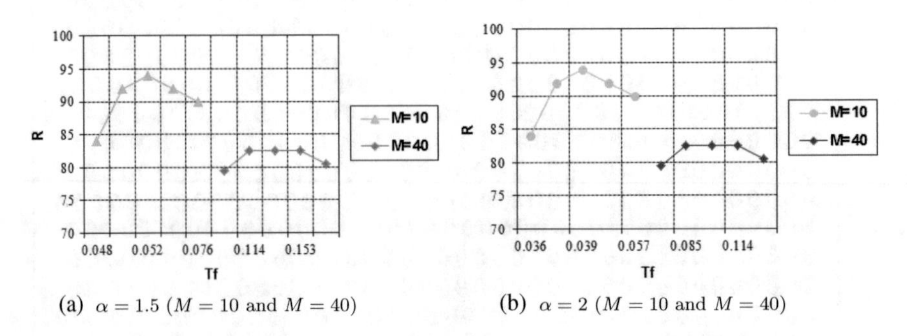

(a) $\alpha = 1.5$ ($M = 10$ and $M = 40$)

(b) $\alpha = 2$ ($M = 10$ and $M = 40$)

Fig. 2.25 The values of T_f as a function of β

The FKCNN is also sensitive to the choice of T_f (less sensitive than sensitive to the choice of β) in the meaning that for α and β fixed, if the value of T_f will be lower than that from Table 2.6 and respectively Table 2.7 then FKCNN will not be able to distinguish between the different images from the training lot.

From Tables 2.6 and 2.7, it results that:

- in case when the values of β decreases and α remains constant, the value of T_f has to decrease in order to achieve good recognition scores on the test lot;
- if we preserve the value of β and diminish the value of α, then T_f has to increase in order that the FKCNN to distinguish the images applied at its input;
- when we experimented FKCNN for a database of 40 classes we have to choose some bigger values of β and T_f than in the situation of working with 10 classes to achieve good recognition rates over the test lot.

The FKCNN approach is a special one as its learning algorithm takes place throughout a single training epoch and its parameters do not need to be refined.

The advantage of using the FKCNN for face recognition consists in the fact that we can obtain good recognition rates over the test lot only through a training epoch, but for the optimum choice of the parameters α, β, and T_f.

2.6.2 Applying Kohonen Maps for Feature Selection

The feature selection stage can be achieved using the Self-organizing Kohonen map (SOKM), which performs [53] the transformation of the original space of the features, in a two-dimensional space, associated to a planar rectangular network.

The projection of the vectors from the initial space is carried out through their association with one of the neurons in the SOKM and the two-dimensional projection will be given by the coordinates of that neuron in the network.

If the training algorithm of the SOKM stops before the network to succeed net classification of the input vectors, then the vectors corresponding to the same class will be associated to some neighboring neurons within the network. The projection of the n-dimensional vectors from the input space into a two-dimensional space is represented by the plane of the network neurons and it makes in the following way [53]:

(a) one assigns a coordinate system to the rectangular network;
(b) determines the projection of an input vector like being given by the coordinates of the neuron, which is associated with the respective vector within the network.

The use of this neural network in the case of the projection in the two-dimensional space aims to highlight the performances of Kohonen network in comparison with situations when in the feature selection are used some nonneural projection algorithms.

Unlike the classification of the n-dimensional vectors using the SOKM, the projection of the vectors in the plane has the advantage that it is much faster (it does not

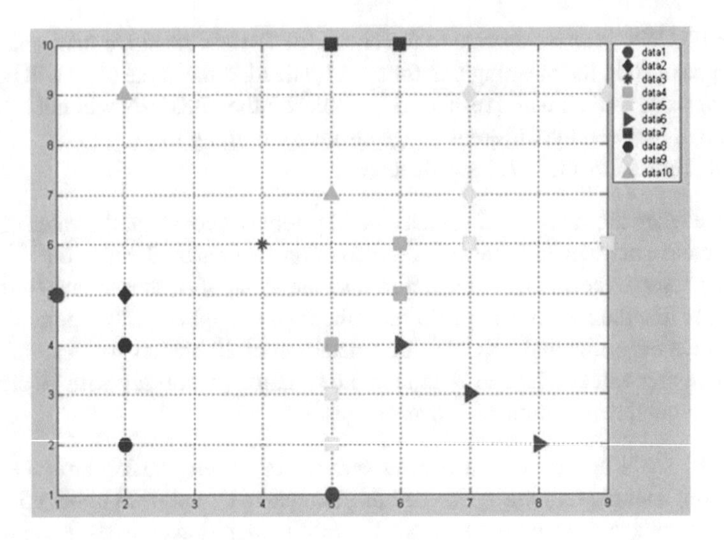

Fig. 2.26 Feature selection using SOKM

```
z1 =1.0e-004 *
[ 0.34127463131750   0.08155682566124   0.00133480036912   0.00082444703065   0.00049793324741
0.00033359878804   0.00023922851977   0.00018001352020   0.00014040657192   0.00011260328114
0.00009233265680  0.00007709623967  0.00006535271130  0.00005610904324]
```

Fig. 2.27 Change of the weights associated with the network neurons

require a large number of iterations) and the resulting two-dimensional vectors can be further classified using a simpler algorithm.

We built a Matlab program to illustrate the training of the SOKM using the 50 images from the ORL face database (see Fig. 2.19) as its input patterns. We have chosen that the value of the neighborhood radius is equal to 10 at the beginning of the refining; after a fixed number of 14 training epochs we represented in Fig. 2.26 the projection result.

We computed

$$z1_{it} = \min_{i=1}^{112\times 92} \|w_{ijk}(it+1) - w_{ijk}(it)\|, (\forall)\, j,k = \overline{1, 10}$$

to depict (in Fig. 2.27) the change of the weights associated with the network neurons, from one epoch it to another $it + 1$.

References

1. I. Iatan and M. Worring. A fuzzy Kwan–Cai neural network for determining image similarity. *BioSystems (Under Review)*, 2016.
2. G. P. Nguyen and M. Worring. Interactive access to large image collections using similarity-based visualization. *Journal of Visual Languages and Computing*, 19:203–224, 2008.
3. G. P. Nguyen and M. Worring. Optimization of interactive visual-similarity-based search. *ACM Transactions on Multimedia Computing Communications and Applications*, 4 (1):1–23, 2008.
4. G. P. Nguyen, M. Worring, and A. W. M. Smeulders. Similarity learning via dissimilarity space in CBIR. In *Proceedings of the ACM SIGMM International Workshop on Multimedia Information Retrieval*, pages 107–116, 2006.
5. G. Strong and M. Gong. Similarity-based image organization and browsing using multi-resolution self-organizing map. *Image and Vision Computing*, 29:774–786, 2011.
6. S.S. Chowhan. Iris recognition using fuzzy min-max neural network. *International Journal of Computer and Electrical Engineering*, 3 (5):743–747, 2011.
7. M. Hariri, S.B. Shokouhi, and N. Mozayani. An improved fuzzy neural network for solving uncertainty in pattern classification and identification. *Iranian Journal of Electrical & Electronic Engineering*, 4 (3):79–93, 2008.
8. H. K. Kwan and Y. Cai. A fuzzy neural network and its application to pattern recognition. *IEEE Trans. on Fuzzy Systems*, 2 (3):185–193, 1997.
9. M. Everingham, L. Van Gool, C. K. I. Williams, J. Winn, and A. Zisserman. A fuzzy neural network and its application to pattern recognition. *IEEE Trans. on Fuzzy Systems*, 88:303–338, 2010.
10. B. Zaka. Theory and applications of similarity detection techniques. http://www.iicm.tugraz.at/thesis/bilal_dissertation.pdf, 2009.
11. K. Suzuki, H. Yamada, and S. Hashimoto. A similarity-based neural network for facial expression analysis. *Pattern Recognition Letters*, 28:1104–1111, 2007.
12. C.M. Hwang, M.S. Yang, W.L. Hung, and M.G. Lee. A similarity measure of intuitionistic fuzzy sets based on the Sugeno integral with its application to pattern recognition. *Information Sciences*, 189:93–109, 2012.
13. Y. Chen, E.K. Garcia, M.Y. Gupta, A. Rahimi, and A. Cazzanti. Similarity-based classification: Concepts and algorithms. *Journal of Machine Learning Research*, 10:747–776, 2009.
14. S. Wang, Q. Huang, S. Jiang, Q. Tian, and L. Qin. Nearest-neighbor method using multiple neighborhood similarities for social media data mining. *Neurocomputing*, 2012.
15. S. Santini and R. Jain. Similarity measures. *IEEE Transactions on Pattern Analysis and Machine Intelligence*, 21 (9):871–883, 1999.
16. G.D. Guo, A.N. Jain, and W.Y. Ma. Learning similarity measure for natural image retrieval with relevance feedback. *IEEE Transactions on Neural Networks*, 13 (4):811–820, 2002.
17. A. Mellouk and A. Chebira. *Machine Learning*. InTech, 2009.
18. T. Mensink, J. Verbeek, F. Perronnin, and G. Csurka. Metric learning for large scale image classification: Generalizing to new classes at near-zero cost. pages 1–14. ECCV - European Conference on Computer Vision, http://hal.inria.fr/docs/00/72/23/13/PDF/mensink12eccv.final.pdf, 2012.
19. N. A. Chinchor, and Wong P. C. Thomas, J. J., M. G. Christel, and W. Ribarsky. Multimedia analysis + visual analytics = multimedia analytics. *Computer Graphics and Applications, IEEE*, 30 (5):52–60, 2010.
20. W. Lin, D. Tao, J. Kacprzyk, Z. Li, E. Izquierdo, and H. Wang. *Multimedia Analysis, Processing and Communications*. Springer-Verlag Berlin Heidelberg, 2011.
21. X. Li, C. G. M. Snoek, and M. Worring. Learning tag relevance by neighbor voting for social image retrieval. In *Proceedings of the 1st ACM international conference on Multimedia information retrieval*, pages 180–187, 2008.
22. J. Perkiö, A. Tuominen, and P. Myllymäki. Image similarity: From syntax to weak semantics using multimodal features with application to multimedia retrieval. *Multimedia Information Networking and Security*, 1:213–219, 2009.

23. A. Huang. Similarity measures for text document clustering. In *NZCSRSC*, pages 49–56, 2008.
24. V. Dutt, V. Chadhury, and I. Khan. Different approaches in pattern recognition. *Computer Science and Engineering*, 1 (2):32–35, 2011.
25. R.C. Chakraborty. Fundamentals of neural networks. http://www.myreaders.info/html/artificial_intelligence.htm, 2010.
26. J.K. Basu, D. Bhattacharyya, and T.H. Kim. Use of artificial neural network in pattern recognition. *International Journal of Software Engineering and Its Applications*, 4 (2):22–34, 2010.
27. Yu J., Wang M., and Tao D. Semisupervised multiview distance metric learning for cartoon synthesis. *IEEE Transactions on Image Processing*, 21 (11):4636–4648, 2012.
28. Yu J., Tao D., and Wang M. Adaptive hypergraph learning and its application in image classification. *IEEE Transactions on Image Processing*, 21 (7):3262–3272, 2012.
29. Yu J., Rui Y., Tang Y.Y., and Tao D. High-order distance-based multiview stochastic learning in image classification. *IEEE Transactions on Cybernetics*, 44 (12):2431–2442, 2014.
30. V. Neagoe and I. Iatan. A neuro-fuzzy approach to face recognition. In *Proceedings of 6th World Multiconference on Systemics, Cybernetics and Informatics (SCI 2002), 14-18 July 2002, Orlando, Florida, XIV*, pages 120–125, 2002.
31. V. E. Neagoe and O. Stănăşilă. *Pattern Recognition and Neural Networks (in Romanian)*. Ed. Matrix Rom, Bucharest, 1999.
32. I. Iatan. Iris recognition by non- parametric techniques. In *Proceedings of International Conference Trends and Challenges in Applied Mathematics, June 20-23, Bucharest*, pages 220–223, 2007.
33. A. Krzyzak. Welcome to pattern recognition homepage. Course Web Page: http://www.cs.concordia.ca/~comp473_2/fall2005/Notes.htm, 2005.
34. F. P. Romero, A. Peralta, A. Soto, J. A. Olivas, and J. Serrano-Guerrero. Fuzzy optimized self-organizing maps and their application to document clustering. *Soft Computing*, 14:857–867, 2010.
35. T.N. Yap. Automatic text archiving and retrieval systems using self-organizing kohonen map. In *Natural Language Processing Research Symposium*, pages 20–24, 2004.
36. I. Bose and C. Xi. Applying Kohonen vector quantization networks for profiling customers of mobile telecommunication services. In *The Tenth Pacific Asia Conference on Information Systems (PACIS 2006)*, pages 1513–1526, 2006.
37. M. Ettaouil, Y. Ghanou, K. El Moutaouakil, and M. Lazaar. Image medical compression by a new architecture optimization model for the Kohonen networks. *International Journal of Computer Theory and Engineering*, 3 (2):204– 210, 2011.
38. M. Ettaouil and M. Lazaar. Compression of medical images using improved Kohonen algorithm. *Special Issue of International Journal of Computer Applications on Software Engineering, Databases and Expert Systems SEDEXS*, pages 41– 45, 2012.
39. M. Lange, D. Nebel, and T. Villmann. *Partial Mutual Information for Classification of Gene Expression Data by Learning Vector Quantization*, pages 259– 270. Advances in Self-Organizing Maps and Learning Vector Quantization. Springer, 2014.
40. A.N. Netravali and B.G. Haskell. *Digital Pictures: Representation and Compression*. Springer, 2012.
41. V. Neagoe. A neural approach to compression of hyperspectral remote sensing imagery. In B. Reusch, editor, *Computational Intelligence, Theory and Applications, International Conference, 7th Fuzzy Days, Dortmund, Germany, October 1-3, 2001*, volume 2206 of *Lecture Notes in Computer Science*, pages 436–449, 2001.
42. N.A. AL-Allaf Omaima. Codebook enhancement in vector quantization image compression using backpropagation neural network. *Journal of Applied Sciences*, 11:3152–3160, 2011.
43. R. Lamba and M. Mittal. Image compression using vector quantization algorithms: A review. *International Journal of Advanced Research in Computer Science and Software Engineering*, 3 (6):354–358, 2013.
44. V.E. Neagoe. Pattern recognition and artificial intelligence (in Romanian), lecture notes, Faculty of Electronics, Telecommunications and Information Technology, University Politehnica of Bucharest. 2000.

45. F.J. Janse van Rensburg, J. Treurnicht, and C.J. Fourie. The use of fourier descriptors for object recognition in robotic assembly. In *5th CIRP International Seminar on Intelligent Computation in Manufacturing Engineering*. https://www.academia.edu/754136/The_Use_of_Fourier_Descriptors_for_Object_Recognition_in_Robotic_Assembly, 2006.
46. R.A. Tuduce. *Signal Theory*. Bren, Bucharest, 1998.
47. T. Kohonen. *Self-Organizing Maps*. Berlin: Springer- Verlag, 1995.
48. V. E. Neagoe. Concurrent self-organizing maps for automatic face recognition. In *Proceedings of the 29th International Conference of the Romanian Technical Military Academy, November 15-16, 2001, Bucharest. Romania*, pages 35–40, 2001.
49. K. E. A. van de Sande, T. Gevers, and C. G. M. Snoek. Evaluating color descriptors for object and scene recognition. *IEEE Transactions on Pattern Analysis and Machine Intelligence*, 32 (9):1582–1596, 2010.
50. M.A. Anjum. *Improved Face Recognition using Image Resolution Reduction and Optimization of Feature Vector*. PhD thesis, National University of Sciences and Technology (NUST) Rawalpindi Pakistan, 2008.
51. S. Dhawan and H. Dogra. Feature extraction techniques for face recognition. *International Journal of Engineering, Business and Enterprise Applications*, 2 (1):1–4, 2012.
52. I. Iatan. *Neuro- Fuzzy Systems for Pattern Recognition (in Romanian)*. PhD thesis, Faculty of Electronics, Telecommunications and Information Technology- University Politehnica of Bucharest, PhD supervisor: Prof. dr. Victor Neagoe, 2003.
53. I. Iatan. Unsupervised neural models and their applications for the feature selection and pattern classification (in Romanian). Master's thesis, Faculty of Mathematics and Computer Science-University of Craiova, PhD supervisor: Prof. dr. Victor Neagoe, June 1998.

Chapter 3
Predicting Human Personality from Social Media Using a Fuzzy Neural Network

Recently, ANNs methods have become useful for a wide variety of applications across a lot of disciplines and in particularly for prediction, where highly nonlinear approaches are required [1]. The advantage of neural networks consists [2] in their ability to represent both linear and non-linear relationships and to learn these relationships directly from the data being modelled. Among the statistical techniques that are widely used is the regression method. The multiple regression analysis has the objective to use independent variables whose values are known to predict the single dependent variable.

Our research objective in the third chapter is to compare [3] the predictive ability of multiple regression and fuzzy neural model, on by which a users personality can be accurately predicted through the publicly available information on their Facebook profile. We'll choose [3] to use the Fuzzy Gaussian Neural Network(FGNN) for predicting personality because it handles nonlinearity associated with the data well. The regression in this paper was chosen as an accepted standard of predicting personality in order to evaluate the FGNN. To test the performance of the neuro-fuzzy prediction achieved based on FGNN we shall use the Normalized Root Mean Square Error (NRMSE).

3.1 Classifying Personality Traits

As we increasingly live our lives online, the type of textual sources available for mining useful knowledge is changing and the type of knowledge that we mine from textual sources changes as well. In social media we record many aspects of our activities: expressing opinions, sharing experiences, social activities, entertainment, learning, looking for information. Such textual records can be mined along a range of highly subjective dimensions. Sentiment analysis [4] is just one such text

I.F. Iatan, *Issues in the Use of Neural Networks in Information Retrieval*,
Studies in Computational Intelligence 661, DOI 10.1007/978-3-319-43871-9_3

mining task, which has been used in a range of settings, both of a personal nature [5] and in commercial settings (product reviews, predicting movie success). Detecting frames, i.e., particular perspectives on a topic [6] is another example. In online reputation management [7], one key task is to estimate the possible impact of a social media message on the reputation of an entity, based in part on the subjective aspects conveyed by the message.

Personality, which constitutes [8] a fundamental component of an individual's affective behavior correlates with some real world behaviors relevant to many types of interactions: it is "useful in predicting job satisfaction, professional and romantic relationship success".[1] Personality not only impacts the formation of social relations, but it also correlates [9] with music taste. It even influences different interfaces, namely how people interact online [9].

"The relationship between real world social networks and personality has been usually studied using the most comprehensive, reliable and useful personality test called *The Big Five*".[2] This test consists in a set of personality concepts; "an individual is associated with five scores that correspond to the five main personality traits: *Openness*, *Conscientiousness*, *Extroversion*, *Ageeableness* and *Neuroticism*. Imaginative, spontaneous and adventurous individuals are high in *Openness*. Ambitious, resourceful and persistent individuals are high in *Conscientiousness*. Individuals who are sociable and tend to seek excitement are high in *Extroversion*. Those high in Agreeableness are trusting, altruistic, tender- minded and are motivated to maintain positive relationship with others. Emotionally liable and impulsive individuals are high in *Neuroticism*".[3]

Personal blogs are a "popular way to write freely and express preferences and opinions on anything that is of interest to someone and therefore provide a useful resource for investigating personality". (see footnote 3).

"Social media are platforms that allow common persons to create and publish contents. For example, the two worldwide popular social media websites, Twitter and Facebook demonstrate its explosive growth and profound influence".[4] "Social media are different from traditional media, such as newspaper, books, and television, in that almost anyone can publish and access information inexpensively using social media". (see footnote 4). "But social media and traditional media are not absolutely distinct in the meaning that the major news channels have official accounts on Twitter and Facebook". (see footnote 4).

---

[1] Golbeck, J., and Robles, C., and Edmondson, M., and Turner, K., Predicting Personality from Twitter, IEEE International Conference on Privacy, Security, Risk, and Trust, and IEEE International Conference on Social Computing, 2011, 149–156.

[2] Quercia, D. and Lambiotte, R. and Stillwell, D. and Kosinski, M. and Crowcroft, J., The Personality of Popular Facebook Users, ACM Conference on Computer Supported Cooperative Work (CSCW 2012), Session: Social Network Analysis, 2012, 955–964.

[3] Iacobelli, F., and Gill, A.J., and Nowson, S., and Oberlander, J., Large Scale Personality Classification of Bloggers, ACII, 2011, 2, 568–577.

[4] Yu, S., and Kak, S., A Survey of Prediction Using Social Media, 2012, http://arxiv.org/ftp/arxiv/papers/1203/1203.1647.pdf.

"Twitter differs from Facebook: people can use the two platforms in very different ways, if they choose to. Facebook is a social networking site that generally connects people who already know each other (e.g., friends, family and co-workers). Instead, Twitter is a social media site on which users can see just about anything about anybody, unless they protect their updates, which only a very tiny minority of active users do".[5]

"There are many forms of social media that include blogs, social networking sites, forums, virtual social worlds, collaborative projects, content communities and virtual game worlds"[6]; these provide venues for diverse people to gather [10], communicate [11], make friends and record experiences and feelings.

The task that we are also going to address in this chapter is interesting, for two reasons. It is a particular instance of an increasingly popular text mining task: automatically inferring personality traits from word use. Algorithmically, the task is interesting because of the highly non-linear nature of the data. We can illustrate this by computing the correlation coefficient (introduced by Pearson) among them.

In this chapter we propose a specific neural network for predicting personality, by special type of Fuzzy Gaussian Neural Network (FGNN) understanding that it has so special the connections (between the second and third layers) and the operations with the nodes, too. The FGNN keeps the advantages of the original fuzzy net described by Chen and Teng [12] for identification in control systems: its structure allows us to construct the fuzzy system rule by rule; if the prior knowledge of an expert is available, then we can directly add some rule nodes and term nodes; the number of rules do not increase exponentially with the number of inputs; elimination of redundant nodes rule by rule. As in the case of the other neural networks, the neurons from the first layer of FGNN only transmit the information to the next level. Each neuron of the second layer (the linguistic term layer), resulted by the fuzzification of the first layer neurons performs a Gaussian membership function. The third layer (the rule layer) of the FGNN computes the antecedent matching by the product operation [3]. The last layer of the FGNN (the output layer) performs the defuzzification of its inputs, providing M non-fuzzy outputs.

We show that our proposed neural network achieves excellent prediction performance, with Normalized Root Mean Square Error (NRMSE) scores of less than 0.1. It is shown to outperforms a number of baselines; the NRMSE (over the test lot) is 0.079 in the case of prediction with FGNN, while in the case of using the Multilayer Perceptron (MP) and Multiple Linear Regression Model (MLRM), the NRMSE is 0.198 and respectively 0.188.

The remainder of the chapter is organized as follows. In Sect. 3.2 we discuss related work. Then, in Sect. 3.3 we introduce and analyze our new neural network,

---

[5]Quercia, D. and Kosinski, M. and Stillwell, D. and Crowcroft, J., Our Twitter Profiles, Our Selves: Predicting Personality with Twitter, IEEE International Conference on Privacy, Security, Risk, and Trust, and IEEE International Conference on Social Computing, 2011, 180–185.

[6]Yu, S., and Kak, S., A Survey of Prediction Using Social Media, 2012, http://arxiv.org/ftp/arxiv/papers/1203/1203.1647.pdf.

fuzzy Gaussian neural networks. We follow with an experimental evaluation of the new model on the task of predicting personality in Sect. 3.4 and then we conclude.

3.2 Related Work

3.2.1 Personality and Word Use

Recent studies have typically focused on broad associations between personality and language use in a variety of different contexts, including directed writing assignments (Hirsh and Peterson, 2009; Pennebaker and King, 1999), structured interviews (Fast and Funder, 2008) and naturalistic recordings of day-to-day speech (Mehl, Gosling, and Pennebaker, 2006). The results of such studies have confirmed and extended previous work on personality; for example, studies have consistently identified theoretically predicted correlations between the dimensions of Extraversion and Neuroticism and usage of words related to a variety of positive and negative emotion categories (Hirsh and Peterson, 2009; Lee, Kim, Seo, and Chung, 2007; Pennebaker and King, 1999). Rentfrow and Gosling (2003) is one of many studies that found the personality as a factor which relates to the music an individual prefers to listen to. Jost et al. (2009) also found that the personality type of an individual was able to predict whether they would be more likely to vote for McCain or Obama in 2008. Hall et al. (2009) performed a regression analysis, using the following two regression algorithms: Gaussian Process and ZeroR.

The study of [13] replicated and extended previous associations between personality and language use in a uniquely large sample of blog-derived writing samples. The results underscore the importance of studying the influence of personality on word use at multiple levels of analysis, and provide a novel approach for refining existing categorical word taxonomies and identifying new and unexpected associations with personality.

Reference [14] made a Pearson correlation analysis between subjects' personality scores and each of the features obtained from analyzing their tweets and public account data. There are a number of significant correlations here, however none of them are strong enough to directly predict any personality trait. He described later in [15] the results of predicting personality traits through machine learning.

More recently, [9] studied the relationship between sociometric popularity (number of Facebook contacts) and personality traits on a far larger number of subjects.

3.2.2 Neural Methods

Neural computing is an information processing paradigm that is inspired by biological systems. An artificial neural network (ANN) is an adaptive mathematical model or a computational structure that is designed to simulate a system of biological neurons to transfer information from its input to output in a desired way. An ANN consists of a large number of interconnecting artificial neurons and employs mathematical or computational models for information processing.

ANN-based methods have been used widely for prediction and classification (Warner and Misra, 1996).

Reference [16] develops a fuzzy neural network approach to financial engineering; this model was successfully applied to the prediction of daily exchange rates (US Dollar-Romanian Lei). In this chapter we extend the application domain of fuzzy neural networks, viz. in the field of text mining, to predict personality traits. This FGNN having M output neurons is unlike the exchange rate FGNN [16], which uses a single neuron in the last layer to estimate the current exchange rate based on the previous m daily exchange rates. For comparison, we considered a linear forecasting approach, corresponding to Higher Order Autoregressive Model (HAR). The aim of the comparison of FGNN versus HAR was [16] to mark both the competition *nonlinear* over *linear* and also that of *neural* over *statistical*.

In 2010 there is [1] a study to compare multilayer perceptron neural networks (NNs) with standard logistic regression (LR). This study is relevant as its aim is to identify key covariates impacting on mortality from cancer causes, disease-free survival (DFS), and disease recurrence using Area Under Receiver-Operating Characteristics (AUROC) in breast cancer patients.

In [2] two techniques for modeling and forecasting the electrical power generated of Nigeria have been analyzed: Neural Network and Statistical Technique. The forecasting ability of the two models is accessed using Mean Absolute Error (MAE), Mean Square Error (MSE), and Root Mean Square Error (RMSE). They have proved the fact that Neural Networks outperform Statistical technique in forecasting. Statistical technique is well established, however their forecasting ability is reduced as the data becomes more complex.

3.3 A Neural Network for Predicting Personality

Before we introduce our fuzzy Gaussian neural network, we recall some background performing regression using neural networks

Statistical methods such as regression analysis, multivariate analysis, Bayesian theory, pattern recognition and least square approximation models have been applied to a wide range of decisions in many disciplines. A multiple linear regression model has used to predict the personality through the publicly available information on their Facebook profile [15].

The ability to predict personality has implications in many areas. Existing research has shown connections between personality traits and success in both professional and personal relationships. If a user's personality can be predicted from their social media profile, online marketing and applications can use this to personalize their message and its presentation.

Artificial Neural Network (ANN) is widely used in various branches of engineering and science and their unique proper of being able to approximate complex and nonlinear equations makes it a useful tool in quantitative analysis. The true power and advantage of neural networks lies in their ability to represent both linear and non-linear relationships and in their being modeled.

We'll choose to use the Fuzzy Gaussian Neural Network for predicting personality because it handles well the nonlinearity associated with the data.

3.3.1 Regression Using Neural Networks

Among the approaches of pattern recognition, the statistical approach has been most intensively studied and used in practice. However, the theory of artificial neural network techniques has been getting significant importance. "The design of a recognition system requires careful attention to the following issues: definition of pattern classes, sensing environment, pattern representation, feature extraction and selection, cluster analysis, classifier design and learning, selection of training and test samples, and performance evaluation".[7]

The statistical methods like regression analysis, multivariate analysis, Bayesian theory, pattern recognition and least square approximation approaches have been applied [17] to make decisions in a lot of disciplines.

"The science of learning plays a key role in the fields of statistics, data mining and artificial intelligence, intersecting with areas of engineering and other disciplines".[8]

There are [18] two stages in the operation of a learning system:

(1) Learning/estimation (from training samples);
(2) Operation/prediction, when predictions are made for future or test samples.

There is [19] the following convention for the prediction tasks:

- *regression* when we predict quantitative outputs;
- *classification* when we predict qualitative outputs.

The Nonlinear Regression Model (NRM) is a generalization of the linear regression model, which is often used for building a purely empirical model. The NRM usually arises when there are physical reasons for believing that the relationship between the response and the predictors follows a particular functional form; it has the form:

$$y_j = \hat{y}_j + \varepsilon_j = \alpha_0 + f(x_{1j}, x_{2j}, \ldots, x_{mj}; \alpha_1, \alpha_2, \ldots, \alpha_m) + \varepsilon_j \tag{3.1}$$

[7]Basu, J.K., and Bhattacharyya, D., and Kim, T.H., Use of Artificial Neural Network in Pattern Recognition, International Journal of Software Engineering and Its Applications, 2010, 4(2), 22–34.

[8]Hastie, T., and Tibshirani, R., and Friedman, J., The Elements of Statistical Learning. Data Mining, Inference, and Prediction, Springer-Verlag Berlin Heidelberg, 2009.

or the vector form [20]:

$$y = F(X, \alpha) + \varepsilon, \tag{3.2}$$

where F is a matrix function of the matrix X and the parameter vector α.

The nonlinear regression (3.2) is nonlinear both with respect to its argument X, and with respect to the vector of regression coefficients α.

The Artificial Neural Networks (ANNs) are well-suited for a very broad class of nonlinear approximations and mappings. The ANN with nonlinear activation functions are more effective than linear regression models in dealing with nonlinear relationships.

A feed-forward neural network is a nonparametric statistical model for extracting nonlinear relations in the data, namely it is a useful statistical tool for nonparametric regression. A two-layer feed-forward neural network with an identity activation function is identical to a linear regression model. The input neurons are equivalent to independent variables or regressors, while the output neuron is the dependent variable. The various weights of the network are equivalent to the estimated coefficients of a regression model. The error term ε is omitted as only the mathematical expression of the computed output value, i.e. the fit is being provided.

A common neural network model configuration is to introduce a layer of hidden neurons between the input and output variables (also called neurons) (see Fig. 3.1).

To truly exploit the potential of neural networks, it is necessary to use a nonlinear activation function. If we are forecasting a variable that may take negative values, it is better to use the hyperbolic tangent as an activation function as it bounded between -1 and 1:

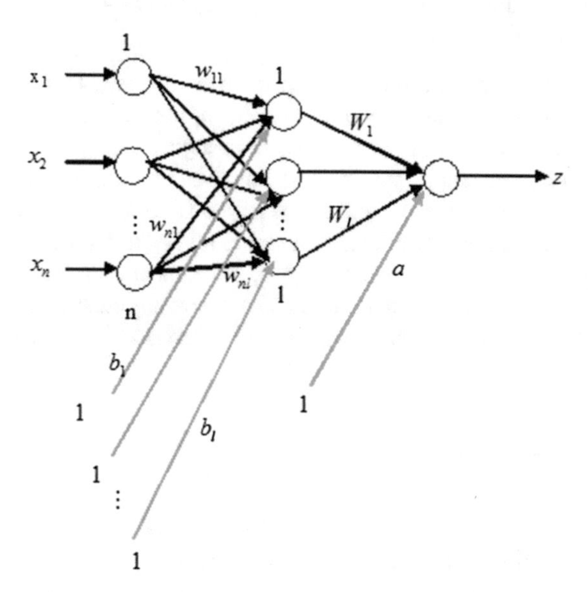

Fig. 3.1 Mathematical structure of a feed-forward neural network with one hidden layer

$$f(x) = \tanh(x) = \frac{e^x - e^{-x}}{e^x + e^{-x}}.$$

The output value of the j-th hidden neuron will be

$$y_j = \tanh\left(\sum_{i=1}^{n} w_{ij}x_i + b_j\right), \ (\forall)\ j = \overline{1, l}, \tag{3.3}$$

where:

- n is the number of the neurons from the input layer;
- l means the number of the neurons from the hidden layer;
- x_i is the ith input;
- w_{ij} are the weight parameters, denoting the weight for the connection linking input i to the hidden unit j;
- b_j means the bias parameters for the hidden units.

The output neuron is given by

$$z = \sum_{j=1}^{l} W_j y_j + a, \tag{3.4}$$

where:

- W_j are the associated weights to the output layer;
- y_j is the output value of the j-th hidden neuron, from (3.4);
- a is a bias term for the output unit.

The parameters b_j and a can also be regarded as the weights for constant inputs of value 1. The optimal parameters $w_{ij},\ W_j,\ b_j,\ a$ for the network are determined after the training algorithm (a back-propagation algorithm), which has the aim to adjust these parameters such that the cost function (it measures the mean square error between the model output z and the observed value z_{obs}) is minimized.

Some advanced neural network techniques are related to more complex statistical methods such as kernel discriminant analysis, k-means cluster analysis or principal component analysis. Some neural networks do not have any close parallel in statistics, such as Kohonen's self-organizing maps, Fuzzy Gaussian Neural Network.

3.3.2 *Fuzzy Gaussian Neural Network*

The Fuzzy Gaussian Neural Network (FGNN) is a special type of neural network, by special type of FGNN understanding that it has so special the connections (between the second and third layers) and the operations with the nodes, too. It represents a modified version of Chen and Teng fuzzy neural network [12], by transforming the function of approximation into a function of classification. The change affects only the equations of the fourth layer, but the structure diagram is similar.

3.3.3 Architecture

The four-layer structure of the Fuzzy Gaussian Neural Network (FGNN) is shown in the Fig. 3.2.

- $X = (x_1, \ldots, x_m)$ represents the vector which one applies to the FGNN input, m being the number of the neurons corresponding to the input layer;
- $\{W_{ij}^3\}_{i=\overline{1,m},\ j=\overline{1,K}}$ is the weight between the $(i-1)K + j$-th neuron of the second layer and the neuron j of the third layer, where K is the number of the neurons from the third layer;
- $\{W_{ij}^4\}_{i=\overline{1,K},\ j=\overline{1,m}}$ is the connection from the neuron i from the third layer and the neuron j from the last layer of the FGNN;
- $Y = (y_1, \ldots, y_M)$ is the output of the FGNN, M meaning the number of classes.

The FGNN keeps the advantages of the original fuzzy net described by Chen and Teng [12] for identification in control systems:

(a) its structure allows us to construct the fuzzy system rule by rule;
(b) if the prior knowledge of an expert is available, then we can directly add some rule nodes and term nodes;
(c) the number of rules do not increase exponentially with the number of inputs;

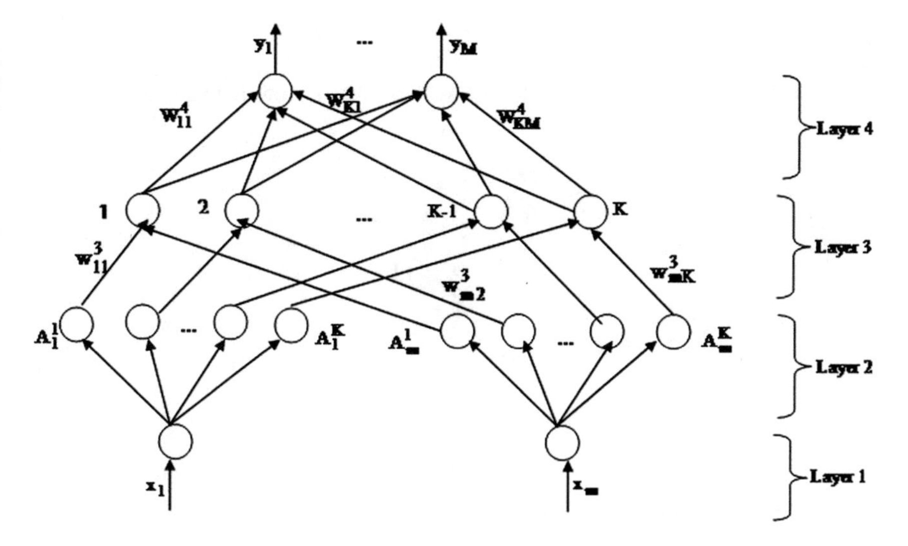

Fig. 3.2 Structure of FGNN

(d) elimination of redundant nodes rule by rule.

The construction of FGNN is based on fuzzy rules of the form:
$\Re_j$: If x_1 is A_1^j and x_2 is A_2^j and ... and x_m is A_m^j
then y_1 is β_1^j and y_2 is β_2^j and ... and y_M is β_M^j,
where:

- m is the dimension of the input vectors (number of the retained features),
- j, $j = \overline{1, K}$ is the rule index,
- M is the number of the output neurons (it corresponds to the number of classes),
- $X = (x_1, \ldots, x_m)^t$ is the input vector, corresponding to the rule R_j,
- A_i^j, $i = \overline{1, m}$ are some fuzzy sets corresponding to the input vector,
- $Y = (y_1, \ldots, y_M)^t$ is the vector of the real outputs, corresponding to the rule R_j,
- β_i^j, $i = \overline{1, M}$ are some fuzzy sets corresponding to the output vector.

The j-th fuzzy rule is illustrated in Fig. 3.3.

FGNN consists in four layers of neurons and each of its neuron performs two actions using two different functions:

1. the first function is the aggregation function $g^k()$, which computes the net input

$$\text{Netinput} = g^k(x^k; W^k), \tag{3.5}$$

   where: the superscript indicates the layer number, therefore $k = \overline{1, 4}$, x^k is input vector of the k-th layer, W^k represents weight vector corresponding to the k-th layer of FGNN;
2. the second function is the nonlinear activation function, denoted W^k, which gives the output:

$$\text{Output} = O_i^k = f^k(g^k), \tag{3.6}$$

   where O_i^k is the i-th output of the respective neuron, from the layer k.

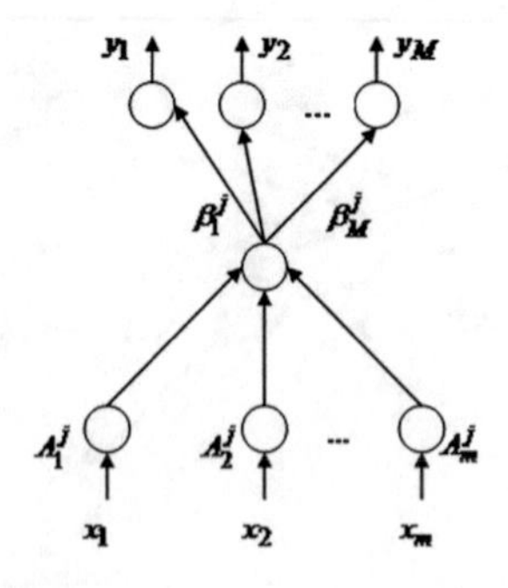

Fig. 3.3 The j-th component of FGNN

As in the case of the other neural networks, the FGNN input layer is a transparent layer, without a role in the data processing, the neurons of the first layer only transmit the information to the next level.

The neurons of the second layer (the linguistic term layer) of the FGNN are membership neurons, resulting by the fuzzification of the first layer neurons. Each neuron of this level performs a Gaussian membership function. The number of neurons characterizing this level is $m \times K$. Each input is transformed by this layer into a fuzzy membership degree.

The third layer of the FGNN is called the rule layer. The connections between the membership neurons of the second layer and the rule neurons that characterize the third layer of the FGNN indicate the premise of the fuzzy rules. This layer computes the antecedent matching by the product operation [3].

The last layer of the FGNN is the output layer, which contains the output neurons. The conclusion (the consequence) of the rules is evidenced by the connections between the neurons of the third layer and the neurons of the output layer. This level performs the defuzzification of its inputs, providing M non-fuzzy outputs.

The FGNN parameters have a physical significance, in the meaning that $m_{ij}, i = \overline{1, m},\ j = \overline{1, K}$ represents the average and $\sigma_{ij}, i = \overline{1, m},\ j = \overline{1, K}$ is the variance of the membership functions corresponding to some fuzzy sets, m being the number of the neurons from the input layer of the FGNN and K representing the number of the fuzzy considered rules.

These parameters of the FGNN one initialize according to an on- line initialization algorithm [21, 22] and they will be refined during the training algorithm.

The training algorithm is of type back- propagation (BP), in order to minimize the error function (like in the case of other fuzzy neural networks as the Neuro Fuzzy Perceptron, the Fuzzy Nonlinear Perceptron based on Alpha Level Sets).

The advantage of the FGNN consists in the fact that, for certain values of the overlapping parameters one achieve very good recognition rates of the test lot.

The FGNN disadvantage is that it requires a large number of term neurons (on the second layer), namely $m \times K$ neurons, for m inputs and K fuzzy rules.

3.3.4 Basic Equations

The operations that execute to each layer of our fuzzy neural network will be described.

The neurons of the first level (*input level*) do not process the signals; they only transmit the information to the next level. The output $O^1_{ki} = x^1_{ki},\ i = \overline{1, m}$ represents the input, n is the number of neurons belonging to the first level of FGNN (equal to the dimension of the input space) and k is the index of the input vector ($k = \overline{1, K}$). The corresponding equations are [12]:

$$g^1_{ki}(x^1_{ki}) = x^1_{ki},\ i = \overline{1, m} \tag{3.7}$$

$$O_{ki}^{1} = f_i^{1}(g_{ki}^{1}) = g_{ki}^{1}(x_{ki}^{1}),\ i = \overline{1, m}. \tag{3.8}$$

Each neuron of the *linguistic term layer* (level 2) performs [12] a Gaussian membership function

$$g_{kij}^{1}(x_{ki}^{2}, m_{ij}, \sigma_{ij}) = -\frac{(x_{ki}^{2} - m_{ij})^2}{\sigma_{ij}^{2}},\ i = \overline{1, m},\ j = \overline{1, K}, \tag{3.9}$$

$$O_{kij}^{2} = f_{ij}^{2}(g_{kij}^{2}) = \exp(g_{kij}^{2}) = \exp\left(-\frac{(x_{ki}^{2} - m_{ij})^2}{\sigma_{ij}^{2}}\right), \tag{3.10}$$

where the corresponding weights to be refined m_{ij} and σ_{ij} denote the mean and variance with respect to A_i^j, $(i = \overline{1, m},\ j = \overline{1, K})$. The number of neurons characterizing this level is nK. Each input x_{ki}^2 is transformed by this layer into a fuzzy membership degree.

The third layer (*rule layer*) computes [12] the antecedent matching by the product operation, according to the relations:

$$g_{kj}^{3}(x_{kij}^{3}, W_{ij}^{3}) = \prod_{i=1}^{n} W_{ij}^{3} x_{kij}^{3},\ j = \overline{1, K}, \tag{3.11}$$

$$O_{kj}^{3} = f_j^{3}(g_{kj}^{3}) = g_{kj}^{3}(x_{kij}^{3}, W_{ij}^{3}),\ j = \overline{1, K}, \tag{3.12}$$

where W_{ij}^3 is the connection weight between the $(i-1)K + j$-th node of the second level and the j-th neuron of the third level, where $i = \overline{1, m},\ j = \overline{1, K}$. We assume $W_{ij}^3 = 1,\ (\forall)\ i = \overline{1, m},\ j = \overline{1, K}$.

The *output level* (fourth level) performs [12] the defuzzification:

$$g_{kj}^{4}(x_{ki}^{4}, W_{ij}^{4}) = \sum_{i=1}^{K} W_{ij}^{4} x_{ki}^{4},\ j = \overline{1, M}. \tag{3.13}$$

We introduce at this level a *sigmoid activation function* in order to apply the FGNN for prediction:

$$O_{kj}^{4} = f_j^{4}(g_{kj}^{4}) = \frac{1}{1 + \exp\left(-\lambda \cdot g_{kj}^{4}(x_{ki}^{4}, W_{ij}^{4})\right)},\ j = \overline{1, M},\ \lambda \in \Re \tag{3.14}$$

where W_{ij}^4 is the connection between the neuron i $(i = \overline{1, K})$ of the third level and the neuron j $(j = \overline{1, M})$ of the fourth level.

3.3.5 On-Line Weight Initialization

The mean initialization algorithm consists in:

$$\begin{cases} m_{ik} = x_{ki}, \ 1 \leq k \leq K-2, \ 1 \leq i \leq m \\ m_{ik} = a_i, \ k = K-1, \ 1 \leq i \leq m \\ m_{ik} = b_i, \ k = K-2, \ 1 \leq i \leq m \end{cases} \tag{3.15}$$

where

$$a_i = \min_{k=1}^{K} x_{ki} \tag{3.16}$$

and

$$b_i = \max_{k=1}^{K} x_{ki}. \tag{3.17}$$

The variance initialization algorithm is:

$$\sigma_{ik} = \frac{\max\{|m_{ik} - m_{iR}|, \ |m_{ik} - m_{iL}|\}}{\sqrt{|\ln \alpha_i|}}, \tag{3.18}$$

$i = \overline{1,m}, \ k = \overline{1,K}$, where:

- $\alpha_i, i = \overline{1,m}, 0 < \alpha_i < 1$ are overlapping factors,
- A_i^R is the closest fuzzy set of A_i^k, on the right of A_i^k,
- A_i^L is the closest fuzzy set of A_i^k, on the left side of A_i^k,
- m_{iR} is the center of A_i^R,
- m_{iL} is the center of A_i^L.

By choosing the means $m_{ik}, \ i = \overline{1,m}, \ k = \overline{1,K}$ using (3.15) and the variances $\sigma_{ik}, \ i = \overline{1,m}, \ k = \overline{1,K}$ based on (3.18), the membership functions of the linguistic labels $A_i^j, \ i = \overline{1,m}, \ k = \overline{1,K}$ satisfy the Theorem 3.1.

Theorem 3.1 ([12]) *Let $A_i = (A_i^1, A_i^2, \ldots, A_i^k)$ be the fuzzy set, in which each linguistic label $A_i^j, \ i = \overline{1,m}, \ k = \overline{1,K}$ has associated a gaussian membership function built on the on-line weight initialization (with the means $m_{ik}, \ i = \overline{1,m}, \ k = \overline{1,K}$ given by (3.15) and the variances $\sigma_{ik}, \ i = \overline{1,m}, \ k = \overline{1,K}$ expressed in (3.18)).*

In this hypotheses, for all $x_i \in X, \ i = \overline{1,m}$ (in the theorem x_i constitutes the arguments of the membership functions, resulted from the respective input vector and X represents the set of the all these arguments) there is $k = \overline{1,K}$ such that $\mu_{A_i^k} \geq \alpha_i$, where $\alpha_i, \ i = \overline{1,m}, \ 0 < \alpha_i < 1$ are the overlapping factors.

3.3.6 Training Algorithm

The training algorithm is of type back- propagation (BP), in order to minimize the error function

$$E = \frac{1}{K}\sum_{k=1}^{K} E_k, \tag{3.19}$$

where

$$E_k = \frac{1}{2}\sum_{i=1}^{M} (d_{ki} - y_{ki})^2\,,\; k = \overline{1, K}, \tag{3.20}$$

represents the error for the rule k.

The weight refinement is given by the gradient rule:

$$\theta_{ij}(t+1) = \theta_{ij}(t) - \eta \cdot \frac{\partial E_k}{\partial \theta_{ij}}, \tag{3.21}$$

where:

- θ_{ij} is the parameter to be refined,
- t is the iteration index,
- η is the learning rate $(0 < \eta < 1)$,
- $d_k = (d_{k1}, \ldots, d_{kM})^t$ is the ideal output vector of the FGNN when at its input is applied the vector having the index k,
- $y_k = (y_{k1}, \ldots, y_{kM})^t$ is the corresponding real output vector of the FGNN ($k = \overline{1, K}$).

The training of this neural network is supervised, namely for of the K vectors from the training lot we know the set of the ideal outputs.

The refining of the FGNN parameters can be divided into two phases, depending on the parameters of premises and respective of conclusions of the rules, as follows:

(a) in the part of the premise of the rules, the means and variances of the Gaussian functions one refine;
(b) in the conclusions of the rules, the weights relating to the latest layer of FGNN must to be refined, the others being equal to 1.

We shall start with the stage (b), from above, namely with the refining of the weights $W_{ij}^4,\ i = \overline{1, K},\ j = \overline{1, M}$, according to the relation:

$$W_{ij}{}^4(t+1) = W_{ij}{}^4(t) - \eta_W \cdot \frac{\partial E_p}{\partial W_{ij}^4}, \tag{3.22}$$

where η_W is the *weight learning rate*.

As

$$\frac{\partial E_p}{\partial W_{ij}^4} = \frac{\partial E_p}{\partial O_{pj}^4} \cdot \frac{\partial O_{pj}^4}{\partial W_{ij}^4} = -(d_{pj} - O_{pj}^4) \cdot \frac{\partial f_j^4}{\partial g_{pj}^4} \cdot \frac{\partial g_{pj}^4}{\partial W_{ij}^4}, \tag{3.23}$$

$$\frac{\partial g_{pj}^4}{\partial W_{ij}^4} = \frac{\partial}{\partial W_{ij}^4}\left(\sum_{i=1}^{K} W_{ij}^4 \cdot x_{pi}^4\right) = x_{pi}^4, \tag{3.24}$$

$$\frac{\partial f_j^4}{\partial g_{pj}^4} = \lambda \cdot O_{pj}^4 \cdot (1 - O_{pj}^4), \tag{3.25}$$

by substituting this relations into (3.22) we achieve the following formula for the refining of the weights W_{ij}^4, afferent to the output layer:

$$W_{ij}^4(t+1) = W_{ij}^4(t) + \eta_W \underbrace{\lambda(d_{pj}^4 - O_{pj}^4)O_{pj}^4(1 - O_{pj}^4)}_{\delta_{pj}^4} x_{pi}^4, \; i = \overline{1, K}, \; j = \overline{1, M}, \tag{3.26}$$

where δ_{pj}^4 means the error term corresponding to the neurons from the output layer.

In the stage (a) we need to refine the means m_{ij} and the variances σ_{ij}, associated to the fuzzy set A_i^j, on the basis of the relations:

$$m_{ij}(t+1) = m_{ij}(t) - \eta_m \cdot \frac{\partial E_p}{\partial m_{ij}}, \; i = \overline{1, m}, \; j = \overline{1, K} \tag{3.27}$$

and respectively

$$\sigma_{ij}(t+1) = \sigma_{ij}(t) - \eta_\sigma \cdot \frac{\partial E_p}{\partial \sigma_{ij}}, \; i = \overline{1, m}, \; j = \overline{1, K}, \tag{3.28}$$

η_m and η_σ being the refining rate of the means and respectively variances, that characterize the Gaussian membership functions O_{ij}^2, $i = \overline{1, m}$, $j = \overline{1, K}$.

We can deduce that:

$$\frac{\partial E_p}{\partial m_{ij}} = \frac{\partial E_p}{\partial g_{pj}^3} \cdot \frac{\partial g_{pj}^3}{\partial O_{pij}^2} \cdot \frac{\partial O_{pij}^2}{\partial m_{ij}} \tag{3.29}$$

$$\frac{\partial E_p}{\partial g_{pj}^3} = \sum_{i=1}^{M} \frac{\partial E_p}{\partial O_{pi}^4} \cdot \frac{\partial O_{pi}^4}{\partial O_{pj}^3} \cdot \frac{\partial O_{pj}^3}{\partial g_{pj}^3} = -\sum_{i=1}^{M}(d_{pi} - O_{pi}^4) \cdot \frac{\partial f_i^4}{\partial g_{pi}^4} \cdot \frac{\partial g_{pi}^4}{\partial O_{pj}^3} \cdot 1,$$

i.e.

$$\frac{\partial E_p}{\partial g^3_{pj}} = \underbrace{-\sum_{i=1}^{M} \delta^4_{pi} \cdot W^4_{ji}}_{\delta^3_{pj}} \tag{3.30}$$

$$\frac{\partial g^3_{pj}}{\partial O^2_{pij}} = \prod_{l=1,\ l\neq i}^{n} O^2_{plj} \tag{3.31}$$

$$\frac{\partial O^2_{pij}}{\partial m_{ij}} = \frac{\partial f^2_{ij}}{\partial g^2_{pij}} \cdot \frac{\partial g^2_{pij}}{\partial m_{ij}} = O^2_{pij} \cdot 2 \cdot \frac{x^2_{pi} - m_{ij}}{(\sigma_{ij})^2} \tag{3.32}$$

Substituting (3.30)–(3.32) into (3.29) it results

$$\frac{\partial E_p}{\partial m_{ij}} = \underbrace{-\ \delta^3_{pj} \cdot \prod_{i=1}^{n} O^2_{pij}}_{\delta^2_{pj}} \cdot \frac{2(x^2_{pi} - m_{ij})}{(\sigma_{ij})^2};$$

hence

$$\frac{\partial E_p}{\partial m_{ij}} = -\delta^2_{pj} \cdot \frac{2(x^2_{pi} - m_{ij})}{(\sigma_{ij})^2}. \tag{3.33}$$

Taking into account of (3.33), the formula (3.27) used to refine the means m_{ij} will become

$$m_{ij}(t+1) = m_{ij}(t) + \eta_m \cdot \delta^2_{pj} \cdot \frac{2(x^2_{pi} - m_{ij})}{(\sigma_{ij})^2},\ i = \overline{1, m},\ j = \overline{1, K}. \tag{3.34}$$

Similarly, we shall have

$$\frac{\partial E_p}{\partial \sigma_{ij}} = \frac{\partial E_p}{\partial g^3_{pj}} \cdot \frac{\partial g^3_{pj}}{\partial O^2_{pij}} \cdot \frac{\partial O^2_{pij}}{\partial \sigma_{ij}} \tag{3.35}$$

$$\frac{\partial O^2_{pij}}{\partial \sigma_{ij}} = \frac{\partial f^2_{ij}}{\partial g^2_{pij}} \cdot \frac{\partial g^2_{pij}}{\partial \sigma_{ij}} = O^2_{pij} \cdot 2 \cdot \frac{(x^2_{pi} - m_{ij})^2}{(\sigma_{ij})^3}. \tag{3.36}$$

By the replacement of the relations (3.30), (3.31) and (3.36) into (3.35) we shall achieve

$$\frac{\partial E_p}{\partial \sigma_{ij}} = \delta^2_{pj} \cdot \frac{2(x^2_{pi} - m_{ij})^2}{(\sigma_{ij})^3}; \tag{3.37}$$

therefore, from (3.28) and (3.37) we can determine the following formula for refining of the variances:

$$\sigma_{ij}(t+1) = \sigma_{ij}(t) + \eta_\sigma \cdot \delta^2_{pj} \cdot \frac{2(x^2_{pi} - m_{ij})^2}{(\sigma_{ij})^3}, \; i = \overline{1, m}, \; j = \overline{1, K}. \tag{3.38}$$

The training algorithm steps are:

Step 1 Apply the vector $X_k = (x_{k1}, \ldots, x_{km})^t$, corresponding to the rule $\Re_k$, to the network input. Initialize the weights W^4_{ij} related to output layer (the only network weights that are not equal to 1, as the weights corresponding to the third layer are equal to 1) with some random values, uniformly distributed in the interval $[-0.5, 0.5]$. Initialize the means m_{ij} and the variances σ_{ij}, $i = \overline{1, m}$, $j = \overline{1, K}$, according to the on-line initialization algorithm.

Step 2 Compute the neuron aggregations in the first layer, using the relation (3.7).

Step 3 Calculate outputs of the neurons from the first layer of FGNN, based on the relation (3.8).

Step 4 Calculate the inputs of neurons in the second layer, using the relation (3.9).

Step 5 Compute the neuron activations in the second layer, using the relation (3.10).

Step 6 Use the relation (3.11) to determine the inputs of the neurons from the third layer.

Step 7 Use the relation (3.12) in order to compute the outputs of the neurons from the third layer.

Step 8 Calculate the neuron aggregations in the output layer, using the relation (3.13).

Step 9 Compute the outputs of the neurons from the fourth layer with (3.14).

Step 10 Calculate the error terms for neurons from the output layer, on the basis of relation:

$$\delta^4_{pj} = \lambda(d^4_{pj} - O^4_{pj}) O^4_{pj} (1 - O^4_{pj}), \; j = \overline{1, M}. \tag{3.39}$$

Step 11 Refine the weights corresponding to the output layer, according to the formula:

$$W^4_{ij}(t+1) = W^4_{ij}(t) + \eta_W \delta^4_{pj} x^4_{pi}, \; i = \overline{1, K}, \; j = \overline{1, M}, \tag{3.40}$$

η_W being a learning rate for the weights.

Step 12 Compute the following error terms:

$$\delta^3_{pj} = -\sum_{i=1}^{M} \delta^4_{pi} W^4_{ji}, \; j = \overline{1, K}. \tag{3.41}$$

Step 13 Calculate the following error terms:

$$\delta_{pj}^{2} = \delta_{pj}^{3} \prod_{i=1}^{n} O_{pij}^{2}, \ j = \overline{1, K}. \tag{3.42}$$

Step 14 Refine the means corresponding to the membership functions O_{pij}^2 with the formula (3.34), η_m being a learning rate for the means.

Step 15 Refine the variances corresponding to the membership functions O_{pij}^2, using the formula (3.38), where η_σ is a learning rate for the variances.

Step 16 Find the error because of the training vector k through the relation (3.20).

Step 17 If $k < K$ (i.e. it isn't traversed the entire training lot) one proceed to the next vector from the training lot and one repeat the algorithm from the *Step 2*.

Step 18 Compute the error corresponding to the respective epoch of the training algorithm, according to the formula (3.19).

Step 19 Test the stop condition of the training algorithm, which is after a fixed number of epochs. If the condition is accomplished, the algorithm stops. Otherwise, we begin a new epoch of learning.

3.4 Experimental Evaluation

3.4.1 Task, Data Set, Data Processing and Evaluation Details

Regression methods analyze [23] relationship between the dependent variable, prediction result, and one or more independent variables, like the social network features. Regression model could be linear and non-linear. One distinguish some methods used in prediction with social media [23]: Bayes classifier, K-nearest neighbor classifier, Artificial Neural Networks, decision trees, model based prediction.

We are trying to find out how relevant is to use the Fuzzy Gaussian Neural Network for predicting personality because it handles well the nonlinearity associated with the data.

We are also asking if the FGNN proves very good prediction performances over a statistical approach of prediction like MLRM and over a neural network as MP, too.

The task that we are addressing is to compare the predictive ability of multiple regression and fuzzy neural model, on by which a users' personality can be accurately predicted through the publicly available information on their Facebook profile. We shall propose to apply the FGNN for predicting a users' Big Five personality traits [13, 14] from the public information they share on Facebook.

The regression in this chapter was chosen as an accepted standard of predicting personality in order to evaluate the FGNN.

We use a data set made available by [13, 14]. The personality test called "The Big Five" (the five factor model of personality) represents the most comprehensive, reliable and useful test of personality concepts. It has emerged as one of the most well-researched and well-regarded measures of personality structure in recent years. This test is used to study the relationship between real world social networks (Twitter or Facebook) and personality.

The Big Five traits are characterized [14] by the following:

- *Openness to Experience*: curious, intelligent, imaginative. High scorers tend to be artistic and sophisticated in taste and appreciate diverse views, ideas, and experiences.
- *Conscientiousness*: responsible, organized, persevering. Conscientious individuals are extremely reliable and tend to be high achievers, hard workers, and planners.
- *Extroversion*: outgoing, amicable, assertive. Friendly and energetic, extroverts draw inspiration from social situations.
- *Agreeableness*: cooperative, helpful, nurturing. People who score high in agreeableness are peace-keepers who are generally optimistic and trusting of others.
- *Neuroticism*: anxious, insecure, sensitive. Neurotics are moody, tense, and easily tipped into experiencing negative emotions.

The data is preprocessed in the following manner: we shall build a data set of 300 vectors, a half of them representing the training lot and the other half being the test lot. These vectors have 20 components, each of them characterizing a personality trait. Each component means a correlation between the Big Five and individual words. For example [13]:

- *Neuroticism* correlates positively with negative emotion words(e.g. awful (0.26), though (0.24), lazy (0.24), worse (0.21), depressing (0.21), irony (0.21), road (−0.2), terrible (0.2), Southern (−0.2), stressful (0.19), horrible (0.19), sort (0.19), visited (−0.19), annoying (0.19), ashamed (0.19), ground (−0.19), ban (0.18), oldest (−0.18), invited (−0.18), completed (−0.18));
- *Extraversion* correlates positively with words reflecting social settings or experiences (e.g. Bar (0.23), other (−0.22), drinks (0.21), restaurant (0.21), dancing (0.2), restaurants (0.2), cats (−0.2), grandfather (0.2), Miami (0.2), countless (0.2), drinking (0.19), shots (0.19), computer (−0.19), girls (0.19), glorious (0.19), minor (−0.19), pool (0.18), crowd (0.18), sang (0.18), grilled (0.18));
- *Openness* shows strong positive correlations with words associated with intellectual or cultural experience (e.g. folk (0.32), humans (0.31), of (0.29), poet (0.29), art (0.29), by (0.28), universe (0.28), poetry (0.28), narrative (0.28), culture (0.28), giveaway (−0.28), century (0.28), sexual (0.27), films (0.27), novel (0.27), decades (0.27), ink (0.27), passage (0.27), literature (0.27), blues (0.26));
- *Agreeableness* correlates with words like: wonderful (0.28), together (0.26), visiting (0.26), morning (0.26), spring (0.25), porn (−0.25), walked (0.23), beautiful (0.23), staying (0.23), felt (0.23), cost (−0.23), share (0.23), gray (0.22), joy (0.22), afternoon (0.22), day (0.22), moments (0.22), hug (0.22), glad (0.22), fuck (−0.22);
- *Conscientiousness* has strong positive correlations with words like: completed (0.25), adventure (0.22), stupid (−0.22), boring (−0.22), adventures (0.2), desperate (−0.2), enjoying (0.2), saying (−0.2), Hawaii (0.19), utter (−0.19), its (−0.19), extreme (−0.19), deck (0.18).

We want to predict M components (M being the number of the neurons from the output layer of FGNN) for every vector in order to complete the behavior corresponding to a person.

For evaluation, we use the Normalized Root Mean Square Error (NRMSE) [25]:

$$\text{NRMSE} = \frac{1}{M}\sqrt{\frac{\sum_{j=m+1}^{m+M}(x_{kj}-\hat{x}_{kj})^2}{\sum_{j=m+1}^{m+M}x_{kj}^2}}. \tag{3.43}$$

Following [26], the prediction is considered: *excellent* if NRMSE ≤ 0.1, *good* if $0.1 <$ NRMSE ≤ 0.2, *fair* if $0.2 <$ NRMSE ≤ 0.3, and *poor* if NRMSE > 0.3.

For significance testing we use the three models: MLRM, MP and FGNN.

3.4.2 Baselines

We use the following baselines: a multiple linear regression model and a multilayer perceptron.

The neural approaches are more efficient than the statistical ones since a feed-forward neural network (NN) is [27] a non-parametric statistical model for extracting nonlinear relations in the data. Neural networks (NNs) are well-suited for a very broad class of nonlinear approximations and mappings.

A Multiple Linear Regression Model (MLRM) can be define as [27]:

$$y_j = \hat{y}_j + \varepsilon_j = \alpha_0 + \sum_{i=1}^{m}\alpha_i x_{ij} + \varepsilon_j,\ (\forall)\ j = \overline{1, K}, \tag{3.44}$$

where:

- $m - 1$ means the prediction order of the model;
- y_j represents the observed value of the predictand;
- K is the number of the multiple predictors x_{ij} for the response variable y_j;
- $\alpha_i,\ (\forall)\ i = \overline{1, m}$ are the regression parameters, α_0 representing a bias term, i.e. an intercept term (a regression constant);
- $\hat{y}_j$ is the predicted value of predictand, namely is the y_j predicted;
- ε_j is the prediction error (error term) in observed value y_j, namely:

$$\varepsilon_j = y_j - \hat{y}_j,\ (\forall)\ j = \overline{1, K}.$$

This type of MLRM can be also written in the vector form [27]:

$$y = X\alpha + \varepsilon, \tag{3.45}$$

where:

$$y = \begin{pmatrix} y_1 \\ \vdots \\ y_K \end{pmatrix}, \quad X = \begin{pmatrix} 1 & x_{11} & \dots & x_{m1} \\ \dots & \dots & \dots & \dots \\ 1 & x_{1K} & \dots & x_{mK} \end{pmatrix}, \quad \alpha = \begin{pmatrix} \alpha_0 \\ \vdots \\ \alpha_m \end{pmatrix}, \quad \varepsilon = \begin{pmatrix} \varepsilon_1 \\ \vdots \\ \varepsilon_K \end{pmatrix}.$$

By finding the optimal values of the parameters α_i, $(\forall)\ i = \overline{1, m}$ the MLRM minimizes the Sum of Squared Errors (SSE):

$$\text{SSE} = \varepsilon^t \varepsilon = (y - X\alpha)^t (y - X\alpha),$$

where the superscript t denotes the transpose.

The vector $\hat{\alpha}$ of the optimal parameters is [27]:

$$\hat{\alpha} = (X^t X)^{-1} X^t y. \tag{3.46}$$

Afterwards, the model (3.44) is estimated by least squares (which yields parameter estimates such that the sum of squares of errors is minimized), the resulting prediction equation is:

$$\hat{y}_j = \hat{\alpha_0} + \sum_{i=1}^{m} \hat{\alpha}_i x_{ij}, \tag{3.47}$$

where "^" denotes the estimated values using the relation (3.46).

A major problem with multiple regression consists in the large number of predictors that are available, although only a few of them are actually significant.

The regression and correlation are related as the both of them are designed to extract linear relations between two variables. In the case of a linear regression model, of the first order, "the slope of the regression line is the correlation coefficient times the ratio of the standard deviation of y to that of x".[9]

In the case of the MLRM applied in [15] for predicting personality, the optimal parameters were computed using the correlations between each profile feature and personality factor.

3.4.3 Experimental Setup

We shall propose to apply the FGNN for predicting a users' Big Five personality traits [13, 14] from the public information they share on Facebook. The Big Five traits are characterized by the following: *Neuroticism*, *Extraversion*, *Openness*, *Agreeableness*, and *Conscientiousness*. There is an extra data set per trait not based on individual words, but on the psychologically defined categories of the Linguistic

[9]Bingham, N.H., and Fry, J.M., Regression. Linear Models in Statistics, Springer, 2010, New York.

Inquiry and Word Count (LIWC) tool; LIWC is the most commonly used language analysis program in studies investigating the relation between word use and psychological variables.

The LIWC 2001 dictionary defines over 70 different categories most of which contain several dozens or hundreds of words. Scores for each category were computed by dividing the number of occurrences of all words within that category by the total number of words in the blog. Detailed descriptions and definitions of the LIWC categories are reported in [13].

We shall normalize the components of a FGNN input vector X_k using a modified formula from [16]:

$$x_{ki} = x_{ki} - \frac{1}{2}\mu_k, \text{ for } k = 1, \ldots, K \text{ and } i = 1, \ldots, m, \tag{3.48}$$

where

$$\mu_k = \frac{1}{m}\sum_{i=1}^{m} x_{ki}, \ (\forall)\ k = \overline{1, K}, \tag{3.49}$$

here, K being the number of the vectors from the input set respectively from the test set. The estimated vector

$$\hat{X}_k = (x_{k1}, \ldots, x_{km}, \hat{x}_{k,m+1}, \hat{x}_{k,m+2}, \ldots, \hat{x}_{k,m+M})$$

for the input vector X_k is determined using the FGNN prediction model according to the formula:

$$\hat{x}_{kj} = O^4_{kj} + \frac{1}{2}\mu_k, \ (\forall)\ k = \overline{1, K}, \ (\forall)\ j = \overline{m+1, m+M}, \tag{3.50}$$

where:

- M is the number of the considered classes,
- $x_{k1}, \ldots, x_{km}$ are the known components corresponding to the the vector X_k,
- $\hat{x}_{k,m+1}, \hat{x}_{k,m+2}, \ldots, \hat{x}_{k,m+M}$ are the predicted components of the vector X_k.

3.4.4 Experimental Results and Analysis

Table 3.1 shows the normalized root mean square errors achieved on the personality trait prediction task, by a multilayer perceptron (MP), a Multiple Linear Regression Model (MLRM), and our fuzzy Gaussian neural network.

Using FGNN we can achieved better results by choosing a different number of neurons on its last layer (see Table 3.2).

Figure 3.4 illustrates how the NRMSE decreases over the test lot during 20 training epochs with FGNN, for different values of M.

Table 3.1 Normalized root mean square errors for personality trait prediction with MP, MLRM, and FGNN

Approach	Performance on	
	Training set	Test set
MP	0.189	0.198
MLRM	0.201	0.188
FGNN	**0.063**	**0.079**

Table 3.2 Normalized root mean square errors for personality trait prediction using FGNN, for different values of M

Number of neurons	Performance on	
	Training set	Test set
8	0.040	0.061
10	0.035	0.047
12	0.030	0.037
15	**0.025**	**0.027**

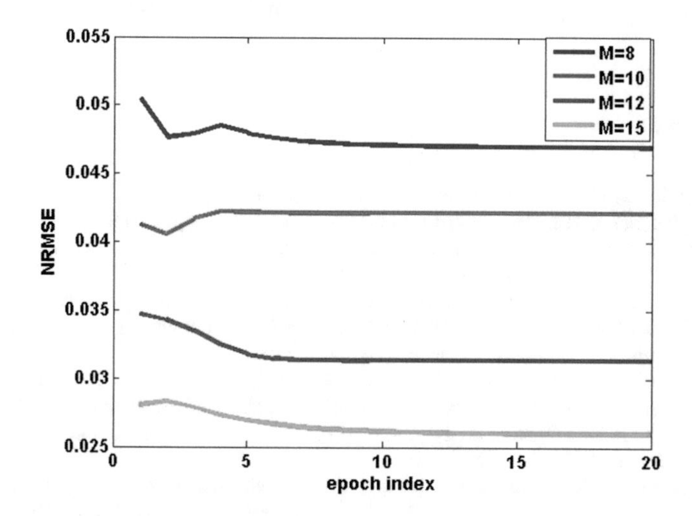

Fig. 3.4 NRMSE over the test lot using FGNN

The three models: MLRM, MP and FGNN have been evaluated using a corresponding test lot, having a number of vectors equal to that of the training lot.

The simulation results given in Tables 3.1 and 3.2 prove very good prediction performances for FGNN over MLRM and MP. The Table 3.2 shows that the number of the neurons from the last layer of FGNN affect the NRMSE: for a bigger number of M, the NRMSE value will decrease. The Fig. 3.4 illustrates this fact proving that for all the considered cases, the NRMSE drops faster during the first 20 epochs and heavier after more training epochs.

The performance of FGNN over MP is based on the fuzzy properties of FGNN, while the MP is a crisp neural network. The comparison of FGNN and respectively MP versus MLRM marks both the competition *nonlinear* over *linear* and of *neural* over *statistical*, too.

With the ability to guess a user's personality traits, many opportunities are opened for personalizing interfaces and information. The ability to predict personality has implications in many areas. Like other studies relating to personality and language we adopted [13] the five factor model of personality, which describes the following traits on a continuous scale: neuroticism, extraversion, openness to experience, agreeableness and conscientiousness.

To emphasize the performances of our proposed approach [3] for predicting personality we have compared it both with a neural method of regression (like MP) and with a nonneural approach (MLRM), too. According with the NRMSE criterion, we have achieved that the prediction with FGNN is better than with others two methods both over the training lot and over the test lot, too.

The next logical step for our work is to improve further the prediction performance of FGNN, for this application and for others from different domains, too.

References

1. V. Bourdès, S. Bonnevay, P. Lisboa, R. Defrance, D. Pérol, S. Chabaud, T. Bachelot, T. Gargi, and S. Nègrier. Comparison of artificial neural network with logistic regression as classificationmodels for variable selection for prediction of breast cancer patient outcomes. *Advances in Artificial Neural Systems*, pages 1–10, 2010.
2. O.S. Maliki, A.O. Agbo, A.O. Maliki, L.M. Ibeh, and C.O. Agwu. Comparison of regression model and artificial neural network model for the prediction of electrical power generated in Nigeria. *Advances in Applied Science Research*, 2(5):329–339, 2011.
3. I. Iatan and M. de Rijke. Predicting Human Personality from Social Media using a Fuzzy Neural Network. *Neural Computing and Applications (Under Review)*, 2016.
4. B. Pang and L. Lee. Opinion mining and sentiment analysis. *Foundations and Trends in Information Retrieval*, 2(1–2):1–135, 2008.
5. M. Kosinski, D. Stillwella, and T. Graepel. Private traits and attributes are predictable from digital records of human behavior. *PNAS*, 2013. http://www.pnas.org/cgi/doi/10.1073/pnas.1218772110.
6. D. Odijk, B. Burscher, R. Vliegenthart, and M. de Rijke. Automatic thematic content analysis: Finding frames in news. In *5th International Conference on Social Informatics (SocInfo2013)*, November 2013.
7. L.A. Fast and D.C. Funder. Personality as manifest in word use: Correlations with selfreport, acquaintance report, and behavior. *J. Pers. Soc. Psychol.*, 94(2):334—346, 2008.
8. F. Iacobelli, A.J. Gill, S. Nowson, and J. Oberlander. Large scale personality classification of bloggers. *ACII*, 2:568–577, 2011.
9. D. Quercia, R. Lambiotte, D. Stillwell, M. Kosinski, and J. Crowcroft. The personality of popular facebook users. In *2012 ACM Conference on Computer Supported Cooperative Work (CSCW 2012), Session: Social Network Analysis*, pages 955–964, 2012.

10. D. Quercia, M. Kosinski, D. Stillwell, and J. Crowcroft. Our twitter profiles, our selves: Predicting personality with twitter. In *2011 IEEE International Conference on Privacy, Security, Risk, and Trust, and IEEE International Conference on Social Computing*, pages 180–185, 2011.
11. S. Asur and B.A. Huberman. Predicting the future with social media. In *Proceedings of the 2010 IEEE/WIC/ACM International Conference on Web Intelligence and Intelligent Agent Technology*, pages 492–499, 2010.
12. C. Y. Chen and C. C. Teng. *Fuzzy Logic and Expert Systems Applications*, chapter Fuzzy Neural Network System in Model Reference Control Systems, pages 285–313. Academic Press, San Diego-Toronto, 1998.
13. T. Yarkoni. Personality in 100,000 words: A large-scale analysis of personality and word use among bloggers. *Journal of Research in Personality*, 44:363–373, 2010.
14. J. Golbeck, C. Robles, M. Edmondson, and K. Turner. Predicting personality from TWITTER. In *IEEE International Conference on Privacy, Security, Risk, and Trust, and IEEE International Conference on Social Computing*, pages 149–156, 2011.
15. J. Golbeck, C. Robles, and K. Turner. Predicting personality with social media. In *Proceedings of alt.chi, ACM Conference on Human Factors in Computing*, pages 253–262, 2012.
16. V. Neagoe, R. Iatan, and I. Iatan. A nonlinear neuro-fuzzy model for prediction of daily exchange rates. In *Proceedings of World Automation Congress WAC'04 Seville, Spain, 17*, pages 573–578, 2004.
17. M.A. Razi and K. Athappilly. A comparative predictive analysis of neural networks (NNs), nonlinear regression and classification and regression tree (cart) models. *Expert Systems with Applications*, 29:65–74, 2005.
18. V. Cherkassky and F. Mulier. *Learning from Data:Concepts, Theory, and Methods*. Wiley-IEEE Press, 2007.
19. T. Hastie, R. Tibshirani, and J. Friedman. *The Elements of Statistical Learning. Data Mining, Inference, and Prediction*. Springer-Verlag Berlin Heidelberg, 2009.
20. G.K. Smyth. *Nonlinear regression. Encyclopedia of Environmetrics*. John Wiley & Sons, Ltd, Chichester, 3, 2002. 1405-1411.
21. V. Neagoe and I. Iatan. Face recognition using a fuzzy-gaussian neural network. In *Proceedings of First IEEE International Conference on Cognitive Informatics, ICCI 2002, 19-20 August 2002, Calgary, Alberta, Canada*, pages 361–368, 2002.
22. V. Neagoe, I. Iatan, and S. Grunwald. A neuro- fuzzy approach to ecg signal classification for ischemic heart disease diagnosis. In *the American Medical Informatics Association Symposium (AMIA 2003), Nov. 8- 12 2003, Washington DC*, pages 494–498, 2003.
23. S. Yu and S. Kak. A survey of prediction using social media. http://arxiv.org/ftp/arxiv/papers/1203/1203.1647.pdf, 2012.
24. T.N. Yap. Automatic text archiving and retrieval systems using self-organizing kohonen map. In *Natural Language Processing Research Symposium*, pages 20–24, 2004.
25. M. Benaddy, M. Wakrim, and S. Aljahdali. Evolutionary prediction for cumulative failure modeling: A comparative study. In *2011 Eighth International Conference on Information Technology: New Generations*, pages 41–47, 2011.
26. H. Zeng, L. Li, J. Hu, L. Liang, J. Li, B. Li, and K. Zhang. Accuracy validation of TRMM multisatellite precipitation analysis daily precipitation products in the lancang river basin of China. *Theoretical and Applied Climatology*, pages 1– 13, 2012.
27. T. Tang. Envrn & geophys data analysis. lecture notes, university of northern bc. http://web.unbc.ca/~ytang/ensc450.html, 2012.

Chapter 4
Modern Neural Methods for Function Approximation

4.1 Mathematical Background

Approximation or representation capabilities of NNs and fuzzy systems have attracted [1] strong research in the past years. Application approaches with their solid results ilustrate that such approximations by the NNs have remarkable accuracy, especially by feedforward neural networks (FNNs) with one hidden layer.

From some literatures it is known [2–7] that the FNNs are characterized as universal approximators.

A typical assertion of such universal approximation have stated as follows: for any given continuous function f defined on a compact set of R^m, there exists a FNN with three layers that approximate the function arbitrarily well. Instead of an explicit formula for the function f, only pairs of input–output data in the form of $(x, f(x))$ are available. Let $x_i \in \mathrm{R}^m$, $i = \overline{1, N}$ be the N input vectors with dimension m and $d_i \in \mathrm{R}$ be a N real number output respectively. We consider an unknown function $f : \mathrm{R}^m \to \mathrm{R}$, that satisfies the interpolation, where

$$f(x_i) = d_i,\ i = \overline{1, N}.$$

The goodness of fit of d_i by the function f, is determined by an error function. A commonly used error function is given by [8]:

$$E(f) = \frac{1}{2}\sum_{i=1}^{N}(d_i - y_i)^2,$$

namely

$$E(f) = \frac{1}{2}\sum_{i=1}^{N}(d_i - f(x_i))^2,$$

where y_i is the actual response.

I.F. Iatan, *Issues in the Use of Neural Networks in Information Retrieval*,
Studies in Computational Intelligence 661, DOI 10.1007/978-3-319-43871-9_4

The main objective of function approximation is [8] to minimize the error function with respect to d_i, $i = 1, ..., N$, namely to enhance the accuracy of the estimation.

An ANN is one of the multiple methods that has been established as a function approximation tool. The classical backpropagation (BP) algorithm is necessary as a gradient-descent optimization method, which adjusts the NN weights to bring its input/output behavior into a desired mapping as of some application environment. The multilayer perceptrons (MLP), along with the backpropagation learning algorithm, are the most popular type of ANN for practical situations. These networks have found their way into countless applications requiring static pattern classification. FNNs such as MLP have been used as an approach to function approximation since they provide a generic black-box functional representation and have been shown to be capable of approximating any continuous function defined on a compact set in R^m with arbitrary accuracy [9], i.e., ANNs as mathematical models are generally enough for most applications; this property is known as *Universal Approximation Property*.

However, BP neural networks have some inherent weaknesses [10]. To resolve such weaknesses of BP neural networks, a Fourier Series Neural Network (FSNN) was built [11, 12] together with its weights-direct determination and structure-growing method.

In the paper [13], we investigate the approximation capabilities of the FSNN to approximate a continuous function. To this end we provide and study a new neural network.

4.1.1 Discrete Fourier Transform

The multidimensional Discrete Fourier Transform (DFT) arises [14] in many fields of science such as image processing, applied physics, mathematics, etc. DFT is a basic operation to transform the ordered sequence of data samples from the time domain into the frequency domain in order to distinguish its frequency components.

A Fourier transform converts a wave in the time domain to the frequency domain. An inverse Fourier transform converts the frequency domain components back into the original time wave.

Because of the periodicity, symmetries, and orthogonality of the basis functions and the special relationship with convolution, the discrete Fourier transform has enormous capacity for improvement of its arithmetic efficiency.

Discrete Fourier transform algorithms include both the direct and the inverse discrete Fourier transform, which converts the result of direct transform back into the time domain.

The Fast Fourier Transform (FFT) is an efficient algorithm for calculating the DFT, which calculates the exact same result (with possible minor differences due to rounding of intermediate results).

The DFT takes a finite sequence of numbers representing one period of $f(U)$ and produces another periodic sequence, with the same period N.

The general form of the s-dimensional FFT from f is defined as follows [14]:

$$F(k_1,\ldots,k_s)=\sum_{u_1=0}^{N_1-1}\sum_{u_2=0}^{N_2-1}\cdots\sum_{u_s=0}^{N_s-1}f(u_1,\ldots,u_s)e^{-j2\pi(k_1u_1/N_1+k_2u_2/N_2+\cdots+k_su_s/N_s)}, \quad (4.1)$$

where N_m is the length of the m-th dimension, $m=\overline{1,s}$ and $f(u_1,\ldots,u_s)$ being the input data.

Conversely, given the DFT of a periodic sequence $f(U)$, denoted $F(K)$, the corresponding time-domain sequence $f(U)$ is obtained from $F(k)$ using the inverse DFT, or IDFT. The Inverse Discrete Fourier Transform (IDFT) can be written [11, 12] as:

$$f(u_1,\ldots,u_s)=\sum_{k_1=0}^{N_1-1}\sum_{k_2=0}^{N_2-1}\cdots\sum_{k_s=0}^{N_s-1}F(k_1,\ldots,k_s)e^{j2\pi(k_1u_1/N_1+k_2u_2/N_2+\cdots+k_su_s/N_s)}. \quad (4.2)$$

In the case when we assume that $N_1=N_2=\cdots=N_m=N$ in (4.1) and respectively in (4.2) we shall have [11, 12]:

$$F(k_1,\ldots,k_s)=\sum_{u_1=0}^{N-1}\sum_{u_2=0}^{N-1}\cdots\sum_{u_s=0}^{N-1}f(u_1,\ldots,u_s)e^{-j2\pi(k_1u_1+k_2u_2+\cdots+k_su_s)/N} \quad (4.3)$$

or

$$F(K)=\sum_{u_1=0}^{N-1}\sum_{u_2=0}^{N-1}\cdots\sum_{u_s=0}^{N-1}f(U)e^{-j2\pi K^TU/N}, \quad (4.4)$$

where: $K=(k_1,\ldots,k_s)^T$, $U=(u_1,\ldots,u_s)^T$ and their components being non-negative integers less than or equal to $N-1$ and:

$$f(u_1,\ldots,u_s)=\frac{1}{N^s}\sum_{k_1=0}^{N-1}\sum_{k_2=0}^{N-1}\cdots\sum_{k_s=0}^{N-1}F(k_1,\ldots,k_s)e^{j2\pi(k_1u_1+k_2u_2+\cdots+k_su_s)/N} \quad (4.5)$$

or

$$f(U)=\sum_{k_1=0}^{N-1}\sum_{k_2=0}^{N-1}\cdots\sum_{k_s=0}^{N-1}F(K)e^{j2\pi K^TU/N}. \quad (4.6)$$

The DFT "may be the most important numerical algorithm in science, engineering, and applied mathematics. Studying the DFT is both valuable in understanding a powerful tool and also a prototype or example of how algorithms can be made efficient and how a theory can be developed to define optimality."[1]

---

[1] Burrus, C.S., and Frigo, M., and Johnson, S.G., and Pueschel, M., and Selesnick, I., Fast Fourier Transforms, 2008, C. Sidney Burrus.

4.1.2 Numerical Methods for Function Approximation

As function approximation plays a very important role in the numerical computation, it represents one of the most important parts of the theory of functions.

In many applications from science and engineering [15], a function of interest is a function of many variables. For instance, the ideal gas law $p = \rho RT$ says [15] that the pressure p is a function of both its density, ρ and its temperature, T, the gas constant R being a material property and not a variable.

In this book, we shall present "how to extend the analysis of functions of a single variable to functions of multiple variables. We shall restrict ourselves with the applications to the case of two variables, i.e. functions of the form $z = z(x; y)$, the extension to larger numbers of variables being relatively straightforward, apart from the fact that functions of three and more variables are somewhat harder to visualize."[2]

There are different methods of approximating functions of many variables: both the classical methods that use Polynomials, Taylor series, or Tensor Products, and the modern methods using Wavelets, Radial Basis Functions, Multivariate Splines, or Ridge Functions.

The theorists in the field of approximation are interested in using Ridge Functions [16] as it constitutes a method for approximating complicated (multivariate) functions by simple functions (linear combinations of Ridge Functions).

A ridge function is a multivariate function of the form [16]:

$$g(\mathbf{a} \cdot \mathbf{x}) = g(a_1 x_1 + \cdots a_n x_n), \tag{4.7}$$

where g is a univariate function and $\mathbf{a} = (a_1, \ldots, a_n)$ is a fixed vector (direction) in $\mathrm{R}^n \backslash \{\mathbf{0}\}$.

The term ridge function is rather recent as certain aspects of the study of ridge functions and various motivations for their research may be found in [16].

The Taylor series expansion of a multivariable function $f(x_1, \ldots, x_n)$ of several variables about a point $\mathbf{a} = (a_1, \ldots, a_n)$ is [15]:

$$f(x_1, \ldots, x_n) = f(a_1, \ldots, a_n) + \left[(x_1 - a_1)\frac{\partial}{\partial x_1} + \cdots + (x_n - a_n)\frac{\partial}{\partial x_n}\right] f(a_1, \ldots, a_n) +$$

$$+\frac{1}{2!}\left[(x_1 - a_1)\frac{\partial}{\partial x_1} + \ldots + (x_n - a_n)\frac{\partial}{\partial x_n}\right]^2 f(a_1, \ldots, a_n) + \cdots +$$

$$+\frac{1}{k!}\left[(x_1 - a_1)\frac{\partial}{\partial x_1} + \cdots + (x_n - a_n)\frac{\partial}{\partial x_n}\right]^k f(a_1, \ldots, a_n) + \cdots$$

and it provides an approximation of the function in the neighborhood of $\mathbf{a} = (a_1, \ldots, a_n)$, which constitutes the center point of Taylor series expansion.

[2]Heil, M., School of Mathematics, Univ. of Manchester, 2010, http://www.maths.manchester.ac.uk/~mheil/Lectures/2M1/Material/Summary.pdf.

The approximation theory [17] is based on the problem of approximating or interpolating a continuous, multivariate function $f : \mathrm{R}^n \to \mathrm{R}^m$ by an approximating function F_W having a fixed number of parameters $W = (w_1, w_2, \ldots, w_p) \in \mathrm{R}^p$. For a choice of a specific approximating function F, it needs to find the set W of parameters "that provides the best possible approximation of f on the set of examples. From this point of view, the approximation of a function is equivalent with the learning problem for a neural network."[3]

4.2 Fourier Series Neural Network (FSNN)

Using NN to implement function approximation shows a new way to the development of function approximation.

The approximation problem can be defined formally as [17].

If $f : X \subseteq \mathrm{R}^n \to \mathrm{R}^m$ is a continuous function, d is an Euclidean distance, and $F_W : \mathrm{R}^n \to \mathrm{R}^m$ is an approximation function that depends continuously on $W \in \mathrm{R}^p$ determine the parameters W^*, such that

$$d(F_{W^*}(x), f(x)) \leq d(F_W(x), f(x)),\ (\forall)\ W \in \mathrm{R}^p. \tag{4.8}$$

A solution of this problem, if it exists, is called [17] a *best approximation.* Typical results deal with the possibility of using a NN to approximate well any continuous function.

The Fourier Series Neural Networks (FSNNs) represent [12] one type of Orthogonal Neural Networks (ONNs) and they are feedforward networks, similarly to sigmoidal neural networks.

The ONNs have many essential advantages, such as [11, 12]

- the learning process convergence fast with the gradient descent algorithm because of the absence of local minima for mean squared error (MSE);
- it is well known both the relation between the number of inputs and outputs and the maximum number of orthogonal neurons, too;
- the NN output is a linear function of weights, that are adjusted during a learning process;
- the initial values of weights aren't so important.

Inputs of FSNN are connected with neurons; these connections are associated with weights. Their values are constant and they don't change during NN training.

Signal from i-th input, multiplied by the aforementioned weights, stimulates $N_i/2 - 1$ neurons in which activation function is the sine, and the same number of neurons in which activation function is cosine, where N_i is a natural even number.

[3]Enăchescu, C., Approximation Capabilities of Neural Networks, Journal of Numerical Analysis, Industrial and Applied Mathematics, 2008, 3(3–4), 221–230.

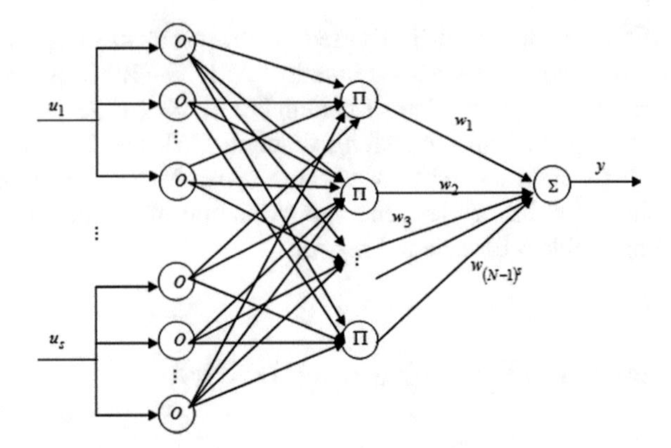

Fig. 4.1 The structure of an orthogonal neural network

The FSNNs are connected with neurons having orthogonal harmonic activation functions from the trigonometric series $c_0,\ c\sin(\alpha u),\ c\cos(\alpha u),\ \alpha = 1, 2, 3\ldots,$ where c_0 and c are constants. These neurons have only a single input.

Multiplying nodes calculate products for all combinations of output signals from the aforementioned neurons. The linear neuron sums the weighted outputs of the multiplying nodes yielding the output of the network. The weights of this neuron are subjected to changes during the network learning process.

The FSNN architecture is designed [11] such that to obtain the same transformation as in a multidimensional Fourier series.

This section describes a method for training a FSNN on the basis of the multidimensional discrete Fourier transform.

Figure 4.1 depicts [11] the ONN with a single output.

In Fig. 4.1, O denotes orthogonal neurons, $\prod$ represents product nodes, and Σ is the linear neuron.

If each input of a FSNN is associated with $N-1$ harmonic neurons, where N is an even number and $c = 1,\ c_0 = 1$, the network output is given by [11]:

$$y = \left(w_1, w_2, \ldots, w_{(N-1)^s}\right) \cdot$$

$$\begin{pmatrix} 1 \\ \sin(u_1) \\ \cos(u_1) \\ \sin(2u_1) \\ \cos(2u_1) \\ \vdots \\ \sin((N/2-1)u_1) \\ \cos((N/2-1)u_1) \end{pmatrix} \otimes \begin{pmatrix} 1 \\ \sin(u_2) \\ \cos(u_2) \\ \sin(2u_2) \\ \cos(2u_2) \\ \vdots \\ \sin((N/2-1)u_2) \\ \cos((N/2-1)u_2) \end{pmatrix} \otimes \cdots \otimes \begin{pmatrix} 1 \\ \sin(u_s) \\ \cos(u_s) \\ \sin(2u_s) \\ \cos(2u_s) \\ \vdots \\ \sin((N/2-1)u_s) \\ \cos((N/2-1)u_s) \end{pmatrix}, \tag{4.9}$$

where:

- $\otimes$ denotes the Kronecker product,
- S is the number of network inputs,
- $w_1, w_2, \ldots, w_{(N-1)^s}$ are network weights,
- $u_1, u_2, \ldots, u_s$ are network inputs,
- $1, \sin(ku_s), cos(ku_s)$ are the activation functions of the neurons connected to the i-th network input, $k = \overline{1, N/2}$,
- y is the network output.

The values we get upon calculating Kronecker's products inside the round brackets in (4.9) are equal to the outputs of multiplying nodes.

The input values to the network must be within the interval $[0, 2\pi)$.

It was found [11, 12] an effective way for allowing to determine the weights $w_1, w_2, \ldots, w_{(N-1)^s}$, which allows to reduce the effect of outliers. The method makes use of s- dimensional DFT, which will be calculated by means of FFT.

Based on Fourier neural network, one can determine [11, 12] a function approximation (an approximated output of the FSNN), which has the following analytical expression:

$$
\begin{aligned}
y \approx r + \frac{2}{N^S} \sum_{k_1=1}^{N/2-1} \sum_{k_2=1}^{N/2-1} \cdots \sum_{k_s=1}^{N/2-1} & \mathrm{Re}\{F(k_1, \ldots, k_s)\} \cos(p(k_1u_1 + k_2u_2 + \cdots + k_su_s)) - \\
& - \mathrm{Im}\{F(k_1, \ldots, k_s)\} \sin(p(k_1u_1 + k_2u_2 + \cdots + k_su_s)) + \\
& + \mathrm{Re}\{F(k_1, \ldots, N - k_s)\} \cos(p(k_1u_1 + k_2u_2 + \cdots + k_{s-1}u_{s-1} - k_su_s)) - \\
& - \mathrm{Im}\{F(k_1, \ldots, N - k_s)\} \sin(p(k_1u_1 + k_2u_2 + \cdots + k_{s-1}u_{s-1} - k_su_s)) + \cdots + \\
& + \mathrm{Re}\{F(k_1, N - k_2, \ldots, N - k_s)\} \cos(p(k_1u_1 - k_2u_2 - \cdots - k_su_s)) - \\
& - \mathrm{Im}\{F(k_1, N - k_2, \ldots, N - k_s)\} \sin(p(k_1u_1 - k_2u_2 - \cdots - k_su_s)),
\end{aligned}
\tag{4.10}
$$

where:

- r is the sum of all combinations of the sums by the type of the previous, where at least one component of the vector K is zero;
- F is the s-dimensional FFT from the function $f(X)$, $X \in \mathrm{R}^s$ (periodic and that satisfies the Dirichlet conditions) to be approximated, namely

$$F(K) = \sum_{u_1=0}^{N-1} \sum_{u_2=0}^{N-1} \cdots \sum_{u_s=0}^{N-1} f(U \cdot p) e^{-j2\pi K^T U/N}; \tag{4.11}$$

- $u_1, u_2, \ldots, u_s$ are the network inputs;

- N is an even number which means the number of points from the data sequence, whose is applied FFT;
- $p = \frac{2\pi}{N}$.

The total computational complexity of the neural method used for the function approximation, on the basis of FSNN is [12]:

$$\mathrm{O}(N^s \log_2 N + (2N)^s).$$

It is much less than that when weights are determined using the least square method.

A drawback of the ONN is caused [11] by the number of weights, which increases exponentially with the dimension of the input data.

4.3 A New Neural Network for Function Approximation

Consider the problem of interpolating a n-variate function $f(x_1, \ldots, x_n)$ on a n-dimensional interval

$$I = \{(x_1, \ldots, x_n) \mid a_i \leq x_i \leq b_i,\ i = \overline{1, n}\},$$

for which we know its values in the points:

$$(x_{11}, x_{12}, \ldots, x_{1m}),\ (x_{21}, x_{22}, \ldots, x_{2m}), \ldots, (x_{n1}, x_{n2}, \ldots, x_{nm})$$

from the neigbourhood of the point $(x_1, \ldots, x_n)$.

In the interpolation or data fitting problem one uncovers some properties satisfied by the function f and then selects an approximant $\hat{f}$ from a family of nice functions that satisfies those properties. The data available about f is often just its value at a set of specified points. The data, however, could include the first or second derivative of f at some of the points. Interpolation methods were historically developed to approximate the value of mathematical and statistical functions from published tables of values.

Modern interpolation theory and practice is concerned with ways to optimally extract data from a function and with computationally efficient methods for constructing and working with its approximant.

As the approximation from (4.10) is best only for sinusoidal or cosinusoidal functions and it doesn't work well for other kind of functions, we shall build an original four layer neural network such as shown in Fig. 4.2.

The first two layers of neurons help us to compute the weight matrix W, which contains the weights between the neurons from the hidden layers.

The weight matrix W can be chosen as follows:

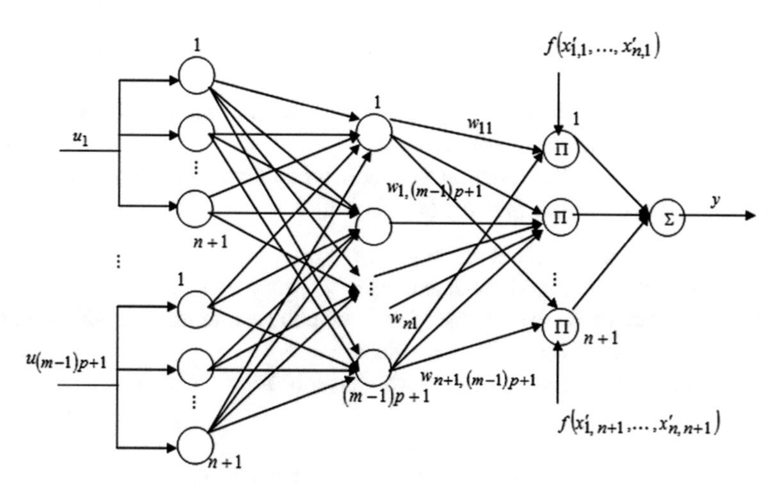

Fig. 4.2 The architecture of our neural network

$$W = \Psi^{+}, \tag{4.12}$$

where Ψ^{+} is the pseudoinverse defined [8] as

$$\Psi^{+} = (\Psi^{T}\Psi)^{-1}\Psi^{T}. \tag{4.13}$$

In the case when one approximates a multivariate polynomial function, which is a generalization of a polynomial to expressions of the form $\sum a_{i_1 i_2 \ldots i_n} \prod_{j=1}^{n} x_j^{i_j}$ we will choose

$$\Psi = \begin{pmatrix} 1 & 0 & 0 & \ldots & 0 \\ 1 & x_{12}-x_{11} & x_{22}-x_{21} & \ldots & x_{n2}-x_{n1} \\ 1 & (x_{12}-x_{11})^2 & (x_{22}-x_{21})^2 & \ldots & (x_{n2}-x_{n1})^2 \\ \ldots & \ldots & \ldots & \ldots & \ldots \\ 1 & (x_{12}-x_{11})^p & (x_{22}-x_{21})^p & \ldots & (x_{n2}-x_{n1})^p \\ 1 & x_{13}-x_{11} & x_{23}-x_{21} & \ldots & x_{n3}-x_{n1} \\ 1 & (x_{13}-x_{11})^2 & (x_{23}-x_{21})^2 & \ldots & (x_{n3}-x_{n1})^2 \\ \ldots & \ldots & \ldots & \ldots & \ldots \\ 1 & (x_{13}-x_{11})^p & (x_{23}-x_{21})^p & \ldots & (x_{n3}-x_{n1})^p \\ \ldots & \ldots & \ldots & \ldots & \ldots \\ 1 & x_{1m}-x_{11} & x_{2m}-x_{21} & \ldots & x_{nm}-x_{n1} \\ 1 & (x_{1m}-x_{11})^2 & (x_{2m}-x_{21})^2 & \ldots & (x_{nm}-x_{n1})^2 \\ \ldots & \ldots & \ldots & \ldots & \ldots \\ 1 & (x_{1m}-x_{11})^p & (x_{2m}-x_{21})^p & \ldots & (x_{nm}-x_{n1})^p \end{pmatrix} \tag{4.14}$$

an invertible matrix, whose columns are a set of linear independent vectors, where $m \geq n + 1,\ p \geq 1$.

Our neural network has $(m - 1)p + 1$ inputs, each of them being a row of the matrix Ψ, namely it has $n + 1$ components.

If $f(x_1, \ldots, x_n)$ is the function that will be approximated then we need the following steps in order to achieve $\hat{f}(x_1, \ldots, x_n)$:

Step 1 Choose a neigborhood of the point $(x_1, \ldots, x_n)$.
Step 2 Determine the pseudoinverse matrix Ψ^+ using (4.13).
Step 3 Substitute in the expression of Ψ^+, the points from the neigborhood of the point $(x_1, \ldots, x_n)$ with their values.
Step 4 Compute the network output, which can be represented as

$$y = \sum_{i=1}^{n+1} \sum_{j=1}^{(m-1)p+1} f(x'_{1i}, x'_{21}, \ldots, x'_{ni}) w_{ij}, \tag{4.15}$$

where

$$(x'_{11}, x'_{21}, \ldots, x'_{n1}), \ldots, (x'_{1,n+1}, x'_{2,n+1}, \ldots, x'_{n,n+1})$$

are the $(n + 1)$ closest neighbours of the point $(x_1, \ldots, x_n)$.

In the case when we need to approximate exponential functions, the matrix Ψ has to be

$$\Psi = \begin{pmatrix} 1 & 1 & 1 & \ldots & 1 \\ 1 & e^{x_{12}-x_{11}} & e^{x_{22}-x_{21}} & \ldots & e^{x_{n2}-x_{n1}} \\ 1 & e^{2(x_{12}-x_{11})} & e^{2(x_{22}-x_{21})} & \ldots & e^{2(x_{n2}-x_{n1})} \\ \ldots & \ldots & \ldots & \ldots & \ldots \\ 1 & e^{p(x_{12}-x_{11})} & e^{p(x_{22}-x_{21})} & \ldots & e^{p(x_{n2}-x_{n1})} \\ 1 & e^{x_{13}-x_{11}} & e^{x_{23}-x_{21}} & \ldots & e^{x_{n3}-x_{n1}} \\ 1 & e^{2(x_{13}-x_{11})} & e^{2(x_{23}-x_{21})} & \ldots & e^{2(x_{n3}-x_{n1})} \\ \ldots & \ldots & \ldots & \ldots & \ldots \\ 1 & e^{p(x_{13}-x_{11})} & e^{p(x_{23}-x_{21})} & \ldots & e^{p(x_{n3}-x_{n1})} \\ \ldots & \ldots & \ldots & \ldots & \ldots \\ 1 & e^{x_{1m}-x_{11}} & e^{x_{2m}-x_{21}} & \ldots & e^{x_{nm}-x_{n1}} \\ 1 & e^{2(x_{1m}-x_{11})} & e^{2(x_{2m}-x_{21})} & \ldots & e^{2(x_{nm}-x_{n1})} \\ \ldots & \ldots & \ldots & \ldots & \ldots \\ 1 & e^{p(x_{1m}-x_{11})} & e^{p(x_{2m}-x_{21})} & \ldots & e^{p(x_{nm}-x_{n1})} \end{pmatrix} \tag{4.16}$$

When you want to approximate sinusoidal or cosinusoidal functions, the matrix Ψ will have the following form:

$$\Psi = \begin{pmatrix} 1 & 0 & 0 & \ldots & 0 \\ 1 & \cos(x_{12}-x_{11}) & \cos(x_{22}-x_{21}) & \ldots & \cos(x_{n2}-x_{n1}) \\ 1 & \cos(2(x_{12}-x_{11})) & \cos(2(x_{22}-x_{21})) & \ldots & \cos(2(x_{n2}-x_{n1})) \\ \ldots & \ldots & \ldots & \ldots & \ldots \\ 1 & \cos(p(x_{12}-x_{11})) & \cos(p(x_{22}-x_{21})) & \ldots & \cos(p(x_{n2}-x_{n1})) \\ 1 & \sin(x_{13}-x_{11}) & \sin(x_{23}-x_{21}) & \ldots & \sin(x_{n3}-x_{n1}) \\ 1 & \sin(2(x_{13}-x_{11})) & \sin(2(x_{23}-x_{21})) & \ldots & \sin(2(x_{n3}-x_{n1})) \\ \ldots & \ldots & \ldots & \ldots & \ldots \\ 1 & \sin(p(x_{13}-x_{11})) & \sin(p(x_{23}-x_{21})) & \ldots & \sin(p(x_{n3}-x_{n1})) \\ \ldots & \ldots & \ldots & \ldots & \ldots \\ 1 & \cos(x_{1,m-1}-x_{11}) & \cos(x_{2,m-1}-x_{21}) & \ldots & \cos(x_{n,m-1}-x_{n1}) \\ 1 & \cos(2(x_{1,m-1}-x_{11})) & \cos(2(x_{2,m-1}-x_{21})) & \ldots & \cos(2(x_{n,m-1}-x_{n1})) \\ \ldots & \ldots & \ldots & \ldots & \ldots \\ 1 & \cos(p(x_{1,m-1}-x_{11})) & \cos(p(x_{2,m-1}-x_{21})) & \ldots & \cos(p(x_{n,m-1}-x_{n1})) \\ 1 & \sin(x_{1m}-x_{11}) & \sin(x_{2m}-x_{21}) & \ldots & \sin(x_{nm}-x_{n1}) \\ 1 & \sin(2(x_{1m}-x_{11})) & \sin(2(x_{2m}-x_{21})) & \ldots & \sin(2(x_{nm}-x_{n1})) \\ \ldots & \ldots & \ldots & \ldots & \ldots \\ 1 & \sin(p(x_{1m}-x_{11})) & \sin(p(x_{2m}-x_{21})) & \ldots & \sin(p(x_{nm}-x_{n1})) \end{pmatrix} \tag{4.17}$$

4.4 Experimental Evaluation

Example 4.1 We shall give an example of using the FSNN to approximate the function $f : \mathrm{R}^2 \rightarrow \mathrm{R}$, defined by

$$f(x, y) = \sin(x^2 + y^2) + \cos(x + y). \tag{4.18}$$

For a two-input network, in this case, using the formula (4.10) we shall have [11]

$$y \approx r + \frac{2}{N^2} \sum_{k_1=1}^{N/2-1} \sum_{k_2=1}^{N/2-1} \mathrm{Re}\{F(k_1, k_2)\} \cos(pk_1u_1 + pk_2u_2) - \mathrm{Im}\{F(k_1, k_2)\} \sin(pk_1u_1 + pk_2u_2) +$$

$$+\mathrm{Re}\{F(k_1, N-k_2)\} \cos(pk_1u_1 - pk_2u_2) - \mathrm{Im}\{F(k_1, N-k_2)\} \sin(pk_1u_1 - pk_2u_2),$$

namely

$$y \approx r + \frac{2}{N^2} \sum_{k_1=1}^{N/2-1} \sum_{k_2=1}^{N/2-1} (\mathrm{Re}\{F(k_1, k_2)\} + \mathrm{Re}\{F(k_1, N-k_2)\}) \cos(pk_1u_1) \cos(pk_2u_2) +$$

$$(-\mathrm{Re}\{F(k_1, k_2)\} + \mathrm{Re}\{F(k_1, N-k_2)\}) \sin(pk_1u_1) \sin(pk_2u_2) +$$

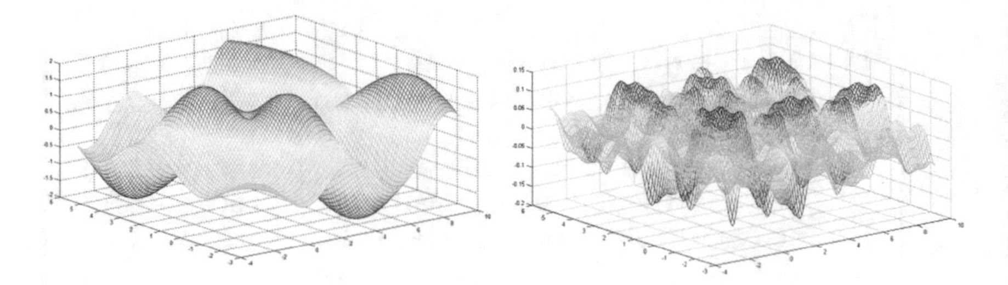

Fig. 4.3 A function and its approximation

$$(-\mathrm{Im}\{F(k_1,k_2)\} - \mathrm{Im}\{F(k_1,N-k_2)\})\sin(pk_1u_1)\cos(pk_2u_2)+$$

$$(-\mathrm{Im}\{F(k_1,k_2)\} + \mathrm{Im}\{F(k_1,N-k_2)\})\cos(pk_1u_1)\sin(pk_2u_2),$$

where

$$r = \frac{F(0,0)}{N^2} + \frac{2}{N^2}\sum_{k_1=1}^{N/2-1}(\mathrm{Re}\{F(k_1,0)\}\cos(pk_1u_1) - \mathrm{Im}\{F(k_1,0)\}\sin(pk_1u_1)) +$$

$$+\frac{2}{N^2}\sum_{k_2=1}^{N/2-1}(\mathrm{Re}\{F(0,k_2)\}\cos(pk_2u_2) - \mathrm{Im}\{F(0,k_2)\}\sin(pk_2u_2)).$$

We shall simulate in MATLAB 7.9 a code in order to use the FSNN for the approximation of the function from (4.18).

The Fig. 4.3 shows both the function and its approximation, in the case when $N = 4$.

To evaluate the approximation results, an error criterion is needed. The following mean square error function is chosen as the error criterion, that is

$$E = \frac{1}{N_1 N_2}\sum_{i=1}^{N_1}\sum_{j=1}^{N_2}(f_{ij} - \hat{f}_{ij})^2, \tag{4.19}$$

where $f_{ij} = f(x_i, y_j)$ and $\hat{f}_{ij} = \hat{f}(x_i, y_j)$, $(\forall)\ i = \overline{1, N_1},\ j = \overline{1, N_2}$.

Applying our neural method in order to approximate the function defined in (4.18), for the following weight matrix

$$\Psi = \begin{pmatrix} 1 & 0 & 0 \\ 1 & \cos(x_2 - x_1) & \cos(y_2 - y_1) \\ 1 & \sin(x_3 - x_1) & \sin(y_3 - y_1) \end{pmatrix} \tag{4.20}$$

we shall obtain an approximation error of 0.0077, with the formula (4.19).

Hence the approximation of this function is finer than that obtained using the FSNN.

If we shall compute the output y of FSNN in order to approximate the value of the function from (4.18) in the point $(6, 4)$ using different values of N we can notice that:

- for $N = 4$ it results $|f(6, 4) - y| = 0.0684$;
- for $N = 6$ it results $|f(6, 4) - y| = 0.0012$.

Using the Taylor polynomial $T_2(x, y)$ of degree 2 about $(5.9, 3.9)$ we obtain $|f(6, 4) - T_2(6, 4)| = 0.65353$, while about $(5.85, 3.85)$: $|f(6, 4) - T_2(6, 4)| = 0.969$; in the case of making the approximation of the function from (4.18) with $T_6(x, y)$ about $(5.85, 3.85)$ we obtain $|f(6, 4) - T_6(6, 4)| = 0.67$. Therefore, the error grows rapidly as (x_0, y_0)-value departs from $(6, 4)$.

Approximating a function with a FSNN is better than with a Taylor series as it has a smaller maximum error and it is more economical.

We can also get a rational function approximation from a FSNN of a function as well.

Example 4.2 We propose to approximate the function as follows:

$$f(x, y) = \frac{x^2}{2 \cdot 3^2} - \frac{y^2}{2 \cdot 2^2}, \tag{4.21}$$

which represents a hyperbolic paraboloid.

We need the following matrix Ψ in order to apply our neural network for the approximation of the function from (4.21):

$$\Psi = \begin{pmatrix} 1 & 0 & 0 \\ 1 & x_2 - x_1 & y_2 - y_1 \\ 1 & x_3 - x_1 & y_3 - y_1 \end{pmatrix} \tag{4.22}$$

Figure 4.4 shows both the hyperbolic paraboloid and its approximation achieved using our neural method.

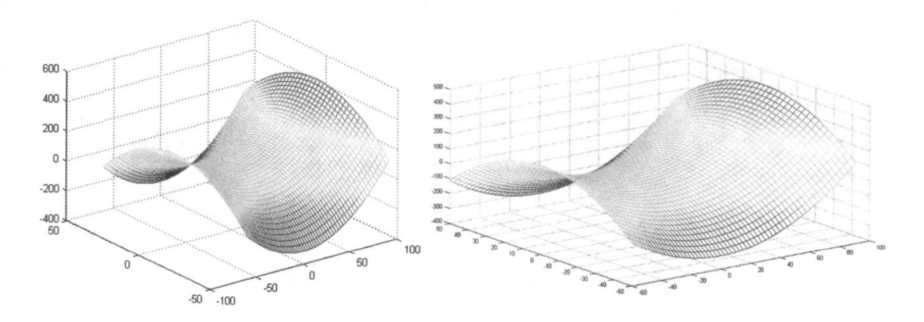

Fig. 4.4 A hyperbolic paraboloid and its approximation

The value of the mean square error function (4.19) corresponding to this approximation is equal to 0.5574.

Using the FSNN, with $N = 4$ for the approximation of the function defined in (4.21), we shall obtain a coarse approximation, as we can notice from Fig. 4.5.

Example 4.3 Let's approximate the function:

$$f(x, y) = \mathrm{e}^{x^2+xy-3y^2}. \tag{4.23}$$

In the aim of approximating the function from (4.23) we shall choose the following matrix Ψ:

$$\Psi = \begin{pmatrix} 1 & 1 & 1 \\ 1 & \mathrm{e}^{x_2-x_1} & \mathrm{e}^{y_2-y_1} \\ 1 & \mathrm{e}^{x_3-x_1} & \mathrm{e}^{y_3-y_1} \end{pmatrix}. \tag{4.24}$$

Figure 4.6 illustrates the function and its approximation achieved using our neural network.

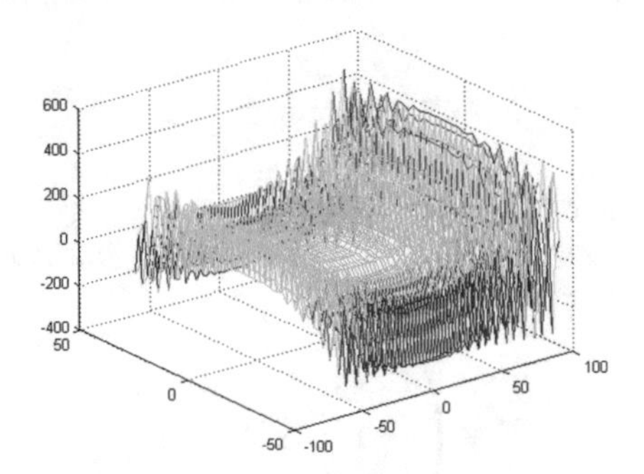

Fig. 4.5 A coarse approximation of the hyperbolic paraboloid

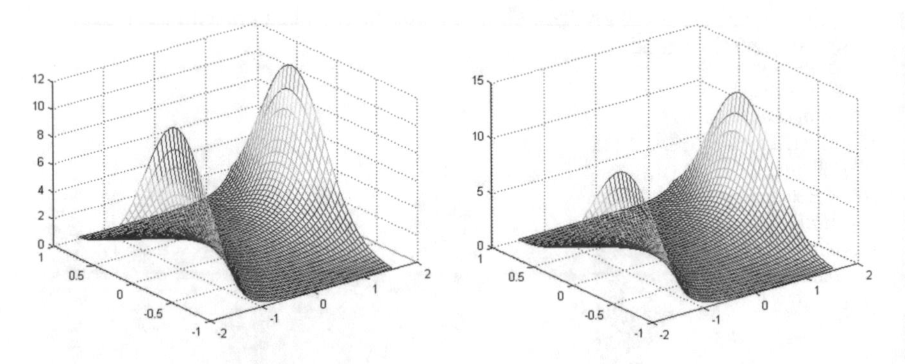

Fig. 4.6 A function and its approximation

In this case, it results a value of the mean square error function corresponding to this approximation equal to 0.1906. If we apply FSNN to approximate the function (4.23) we shall obtain a coarse approximation.

References

1. X.J. Zeng, J.A. Keane, J.Y. Goulermas, and P. Liatsis. Approximation capabilities of hierarchical neural-fuzzy systems for function approximation on discrete spaces. *International Journal of Computational Intelligence Research*, 1:29–41, 2005.
2. G.A. Anastassiou. *Intelligent Systems: Approximation by Artificial Neural Networks*. Berlin: Springer-Verlag, 2011.
3. R. Lovassy, L.T. Kóczy, I.J. Rudas, and L. Gál. Fuzzy neural networks as good function approximators. *Óbuda University e-Bulletin*, 2(1):173–182, 2011.
4. A. Mishra and Z. Zaheeruddin. Design of hybrid fuzzy neural network for function approximation. *Journal of Intelligent Learning Systems and Applications*, 2(2):97–109, 2010.
5. O. Rudenko and O. Bezsonov. Function approximation using robust radial basis function networks. *Journal of Intelligent Learning Systems and Applications*, 3:17–25, 2011.
6. J. Wang and Z. Xu. Neural networks and the best trigomometric approximation. *Journal of Systems Science and Complexity*, 24:401–412, 2011.
7. J. Wang, B.L. Chen, and C.Y. Yang. Approximation of algebraic and trigonometric polynomials by feedforward neural networks. *Neural Computing & Applications*, 21:73–80, 2012.
8. Z. Zainuddin and P. Ong. Function approximation using artificial neural networks. *WSEAS Transactions on Mathematics*, 7(6):333–338, 2008.
9. R. Ratsko, A. Selmic, and Lewis F.L. Neural network approximation of piecewise continuous functions: application to friction compensation. *IEEE Transactions on Neural Networks*, 13(3):745–751, 2002.
10. B. Niu, Y.L. Zhu, K.Y. Hu, S.F. Li, and X.X. He. A cooperative evolutionary system for designing neural networks. In *Lecture Notes in Computer Science*, number 4113, 2006.
11. K. Halawa. Determining the weights of a fourier series neural network on the basis of the multidimensional discrete fourier transform. *International Journal of Applied Mathematics and Computer Science*, 18(3):369–375, 2008.
12. K. Halawa. Fast and robust way of learning the fourier series neural networks on the basis of multidimensional discrete fourier transform. In *Lecture Notes in Computer Science*, number 5097, 2008.
13. I. Iatan, S. Migorski, and M. de Rijke. Modern neural methods for function approximation. *Applied Soft Computing*, (Under review), 2016.
14. Z. Chen and L. Zhang. Vector coding algorithms for multidimensional discrete Fourier transform. *Journal of Computational and Applied Mathematics*, 212:63–74, 2008.
15. M. Heil. School of Mathematics, Univ. of Manchester. http://www.maths.manchester.ac.uk/~mheil/Lectures/2M1/Material/Summary.pdf, 2010.
16. V.E. Ismailov. Representation of multivariate functions by sums of ridge functions. *Journal of Mathematical Analysis and Applications*, 331:184–190, 2007.
17. C. Enăchescu. Approximation capabilities of neural networks. *Journal of Numerical Analysis, Industrial and Applied Mathematics*, 3(3–4):221–230, 2008.

Chapter 5
A Fuzzy Gaussian Clifford Neural Network

5.1 Introduction

A quick tour of relevant algebra is [1]:

Groups→Rings→Fields→Vector Spaces→Algebras→Clifford Algebras.

The geometric interpretations of geometric multiplication originally led W. K. Clifford (a young English Goldsmid professor of applied mathematics at the University College of London) to name these algebras as Geometric Algebras.

Clifford "has unified and generalized in his geometric algebras (=Clifford algebras) the works of Hamilton and Grassmann by finalizing the fundamental concept of directed numbers."[1] An algebra that results [1] from extending a real vector space in this manner is called a normed, associative algebra, or Clifford Algebra. "It unifies all branches of physics, and has found rich applications in robotics, signal processing, ray tracing, virtual reality, computer vision, vector field processing, tracking, geographic information systems and neural computing." (see footnote 1).

Although the algebras of Clifford and Grassmann are well known to pure mathematicians [2], the physicists has abandoned them long ago in the favor of the vector algebra of Gibbs, which is indeed used today in most fields of physics. Clifford algebra which has been developed [3] from Grassmann algebra, is applied in a lot of fields of quantum physics, science calculation.

Neural computation in Clifford algebras, including "familiar complex numbers and quaternions as special cases, has recently become an active research field."[2]

"The Clifford algebra is a powerful new mathematical tool for geometric calculus, it provided an algebraic tools for geometric objects. Moreover, the Clifford

[1]Hitzer, E., Introduction to Clifford's Geometric Algebra, SICE Journal of Control, Measurement, and System Integration, 2011, 4(1), 1–10.

[2]Buchholz, S., and Tachibana, K., and Hitzer, E. M.S., Optimal Learning Rates for Clifford Neurons, Lecture Notes in Computer Science, 2007, 4668, 864–873.

I.F. Iatan, *Issues in the Use of Neural Networks in Information Retrieval*,
Studies in Computational Intelligence 661, DOI 10.1007/978-3-319-43871-9_5

algebra-based method was more reducing the computational complexity of MDFNs (Multi-Degree of Freedom Neurons) than that of Euclidean geometrics."[3]

To establish the theory of Clifford neural computation from the outlined motivation as a powerful model-based approach is the main goal of the paper [4]. We shall start with the design of Clifford neurons for which weight association is interpretable as a geometric transformation. Afterwards, it is proved "how different operation modes of such neurons can be selected by different data representations."[4] From the neuron level we then proceed to the Fuzzy Clifford Gaussian Neural Network.

5.1.1 Basics of Clifford Algebras

We shall define the name of algebra in order to understand what a Clifford algebra is. In mathematics, an algebra over a field F is a vector space equipped with a bilinear vector product.

An algebra that is a finite-dimensional vector space over F is called a finite-dimensional algebra. If the linear space is of finite dimension, say n, then the algebra is also said to be n-dimensional.

"William Hamilton invented quaternion and completed the calculus of quaternions to generalize complex numbers in 4 dimension (one real part and 3 imaginary numbers). Quaternion is a geometrical operator to represent the relationship (relative length and relative orientation) between two vectors in 3D space."[5]

A Clifford algebra is a generalization of the complex numbers C and the Hamilton's quaternions are denoted by H. The quaternions can be written in the vector form [5]:

$$(a, b, c, d) = a + b\mathrm{i} + c\mathrm{j} + d\mathrm{k}, \;\; a, b, c, d \in \Re,$$

i, j, k being the basis vectors of H and they are often named imaginary units; the following relations hold among them [6]:

$$\begin{cases} \mathrm{i}^2 = \mathrm{j}^2 = \mathrm{k}^2 = \mathrm{ijk} = -1 \\ \mathrm{jk} = -\mathrm{kj} = \mathrm{i} \\ \mathrm{ki} = -\mathrm{ik} = \mathrm{j} \\ \mathrm{ij} = -\mathrm{ji} = \mathrm{k}, \end{cases} \tag{5.1}$$

which shows that multiplication in H is not commutative.

---

[3] Ding, L., and Feng, H., and Hua, L., Multi-degree of Freedom Neurons Geometric Computation Based on Clifford Algebra, Fourth International Conference on Intelligent Computation Technology and Automation, 2011, 177–180.

[4] Buchholz, S., A Theory of Neural Computation with Clifford Algebras, Kiel, 2005, http://www.informatik.uni-kiel.de/inf/Sommer/doc/Dissertationen/Sven_Buchholz/diss.pdf.

[5] Ahn, S. H, 2009. Math. http://www.songho.ca/math/index.html.

If $a + bi + cj + dk$ is any quaternion, then a is called its scalar part and $bi + cj + dk$ is called its vector part; they can be used to represent rotations in three-dimensional space. A quaternion whose scalar part is zero is called a pure quaternion. The scalars a, b, c, d are called the components of the quaternion.

Because the vector part of a quaternion is a vector in $\Re^3$, the geometry of $\Re^3$ is reflected in the algebraic structure of the quaternions. The elements i, j, and k will denote both imaginary basis vectors of H and the standard orthonormal basis in $\Re^3$.

Unlike the complex numbers, the quaternions are an example of a noncommutative algebra. The quaternions form a 4-dimensional unitary associative algebra over the reals (but not an algebra over the complex numbers, since if complex numbers are treated as a subset of the quaternions, complex numbers and quaternions do not commute).

The Clifford algebras, which are useful in geometry and physics can be thought of as directed number systems. Because Clifford algebras are vector spaces, a wide variety of results from Linear Algebra can be applied to them. Clifford algebras are generated by quadratic spaces from which they inherit metric structure.

Euler's formula [7] can be used to represent a 2D point with a length and angle on a complex plane. Multiplication of two complex numbers implies a rotation in 2D. One may think instantly it can be extended to 3D rotation by adding additional dimension.

5.1.2 *Generation of Clifford Algebras*

We shall put more emphasis to the direct generation of Clifford algebras by nondegenerate quadratic forms.

So let be Q a non-degenerate quadratic form on $\Re^n$. For shortness, we shall call $(Q, \Re^n)$ a quadratic space. By a theorem of linear algebra (Sylvester's Law of Inertia) there exists a basis of $\Re^n$ such that [8]

$$Q(v) = v_1^2 + \ldots + v_p^2 - v_{p+1}^2 - \ldots v_{p+q}^2, \ (\forall)\ (v_1, v_2, \ldots, v_p) \in \Re^n, \ (\forall) p, q \in \mathrm{N}, \ p + q = n. \tag{5.2}$$

For the vectors of an orthonormal basis $\mathrm{B} = \{e_1, e_2, \ldots, e_n\}$ of $\Re^n$ we get from (5.2):

$$Q(e_i) = \begin{cases} 1, \ i \leq p \\ -1, \ p + 1 \leq i \leq p + q \\ 0, \ p + q + 1 \leq i \leq n \end{cases} \tag{5.3}$$

$$Q(e_i + e_j) - Q(e_i) - Q(e_j) = 0, \ (\forall)\ i, j = \overline{1, n}. \tag{5.4}$$

A so-defined quadratic space is denoted by $\Re^{p,q}$. A Clifford algebra $C_{p,q}$ of $\Re^{p,q}$ is [8] an associative algebra with unity 1 over $\Re^{p,q}$, of dimension 2^{p+q}, containing $\Re^{p,q}$ and $\Re$ as distinct subspaces iff:

(a) $v \otimes_{p,q} v = Q(v),\ (\forall)\ v \in \Re^{p,q}$,

(b) $C_{p,q}$ is generated as an algebra by $\Re^{p,q}$,

(c) $C_{p,q}$ is not generated by any proper subspace of $\Re^{p,q}$.

Associated with every space $\Re^{p,q}$ is a unique Clifford algebra of maximal dimension. Condition (a) in the previous definition is equivalent to

$$e_i^2 = Q(e_i). \tag{5.5}$$

As well as for vector addition and scalar multiplication, we have a noncommutative product which is associative and distributive over addition this is the geometric or Clifford product.

A Clifford algebra $C_{p,q}$ of the quadratic space $\Re^{p,q}$ can be generated using the non-degenerate quadratic form $Q : \Re^n \to \Re$, defined in (5.2), such that

$$v \otimes_{p,q} v = Q(v); \tag{5.6}$$

the multiplication $\otimes_{p,q}$ is called the geometric product.

Examples of Clifford algebras are [5] the real numbers $\Re$ corresponding to $C_{0,0}$ (namely $C_{0,0} \cong \Re$), the complex numbers C corresponding to $C_{0,1}$, $C_{0,1} \cong$ C and the quaternions H corresponding to $C_{0,2}$, respectively ($C_{0,2} \cong$ H).

The bilinear form associated to the quadratic form Q is

$$b(u, v) = u \otimes_{p,q} v,\ (\forall)\ u, v \in \Re^{p,q}. \tag{5.7}$$

We shall have:

$$e_i \otimes_{p,q} e_j + e_j \otimes_{p,q} e_i = 2b(e_i, e_j),\ (\forall)\ i, j = \overline{1, n},\ i \neq j. \tag{5.8}$$

We know that

$$2b(e_i, e_j) = Q(e_i + e_j) - Q(e_i) - Q(e_j),\ (\forall)\ i, j = \overline{1, n},\ i \neq j. \tag{5.9}$$

Taking into account the relation (5.7), from (5.9) it results that

$$2b(e_i, e_j) = 0,$$

namely

$$e_i e_j + e_j e_i = 0,\ (\forall)\ i, j = \overline{1, n},\ i \neq j \tag{5.10}$$

or

$$e_i \otimes_{p,q} e_j = -e_j \otimes_{p,q} e_i = 2b(e_i, e_j),\ (\forall)\ i, j = \overline{1, n},\ i \neq j. \tag{5.11}$$

A further very important consequence of equation (5.11) is, that only Clifford algebras up to dimension 2 are commutative ones.

We shall now give explicitly a basis of a Clifford algebra in terms of the basis vectors $\mathrm{B} = \{e_1, e_2, \ldots, e_n\}$ of the underlying quadratic space.

If $\mathrm{B} = \{e_1, e_2, \ldots, e_n\}$ is the canonical basis of $\Re^n$, then the canonical basis of $C_{p,q}$ consists in the set:

$$\mathrm{B}_1 = \{1, e_{a_1}, e_{a_2}, \ldots, e_{a_r}, e_{a_1}e_{a_2}, \ldots, e_{a_1}e_{a_2}\ldots e_{a_r},\ 1 \leq a_1 \leq a_2 \ldots a_r \leq p+q\};$$

hence the Clifford algebra has the dimension 2^{p+q}. An element of a Clifford algebra is called a multivector, due to the fact that it consists of objects of different types by definition.

An element $v \in C_{p,q}$, which can be written in the following form:

$$v = a_0 + a_1 e_{a_1} + \cdots + a_r e_{a_r} + a_{r+1} e_{a_1} e_{a_2} + \cdots + a_u e_{a_1} e_{a_2} \ldots e_{a_r},\ u = 2^{p+q} - 1 \tag{5.12}$$

is an example by multivector; the 0-vectors are [8] the scalars (a_0 is a 0-vector); the 1-vectors are the vectors ($e_{a_1}, e_{a_2}, \ldots, e_{a_r}$ are some 1-vectors); the 2-vectors are bivectors ($e_{a_1}e_{a_2}, \ldots, e_{a_{r-1}}e_{a_r}$ are some bivectors), the 3-vectors are trivectors ($e_{a_1}e_{a_2}e_{a_3}, \ldots, e_{a_{r-2}}e_{a_{r-1}}e_{a_r}$ are trivectors).

The Hamilton product of the quaternions $z_1 = (a_1, b_1, c_1, d_1)$ and $z_2 = (a_2, b_2, c_2, d_2)$ is accomplished as follows:

Step 1 One assumes the basis $\mathrm{B}' = \{1, e_1, e_2, e_1e_2\}$ corresponding to the Clifford algebra $C_{0,2}$.

Step 2 One applies the formula (5.7) in order to compute: $b(1, 1) = 1$, $b(e_1, e_1) = b(e_2, e_2) = -1$, etc.

The associated matrix to the bilinear form b in the basis B' will be

$$\mathbf{A} = \begin{array}{c} \\ a_1 \\ b_1 \\ c_1 \\ d_1 \end{array} \begin{array}{c} \begin{array}{cccc} a_2 & b_2 & c_2 & d_2 \end{array} \\ \begin{pmatrix} 1 & e_1 & e_2 & e_1e_2 \\ e_1 & -1 & e_1e_2 & -e_2 \\ e_2 & -e_1e_2 & -1 & e_1 \\ e_1e_2 & e_2 & -e1 & -1 \end{pmatrix} \end{array}$$

it results that the expression of the bilinear form b in the basis B' is

$$\begin{aligned} b(z_1, z_2) = {} & a_1a_2 + a_1b_2e_1 + a_1c_2e_2 + a_1d_2e_1e_2 + b_1a_2e_1 - b_1b_2 + b_1c_2e_1e_2 \\ & - b_1d_2e_2 + c_1a_2e_2 - c_1b_2e_1e_2 - c_1c_2 + c_1d_2e_1 + d_1a_2e_1e_2 + d_1b_2e_2 \\ & - d_1c_2e_1 - d_1d_2, \end{aligned}$$

namely

$$z_1 \otimes z_2 =$$

$$(a_1a_2-b_1b_2-c_1c_2-d_1d_2,\ a_1b_2+b_1a_2+c_1d_2-d_1c_2,\ a_1c_2-b_1d_2+c_1a_2+d_1b_2,\ a_1d_2+b_1c_2-c_1b_2+a_2d_1),$$

where:

$$e_1^2 = e_2^2 = (e_1e_2)^2 = -1.$$

The multiplication of quaternions is associative, which can be checked directly.

The multiplicative inverse of any nonzero quaternion $z = a + b\mathrm{i} + c\mathrm{j} + d\mathrm{k},\ a, b, c, d \in \Re$ is given by [7]:

$$z^{-1} = \frac{a - b\mathrm{i} - c\mathrm{j} - d\mathrm{k}}{a^2 + b^2 + c^2 + d^2},$$

which is much easier to verify.

In both C and H a division is defined. If this holds for a general algebra one speaks of a division algebra.

Let $z = q_0 + q = q_0 + q_1\mathrm{i} + q_2\mathrm{j} + q_3\mathrm{k}$ be a quaternion. The complex conjugate of z, denoted z^*, is defined as [7]

$$z^* = q_0 - q = q_0 - q_1\mathrm{i} - q_2\mathrm{j} - q_3\mathrm{k}.$$

The norm of a quaternion z, denoted by $|z|$ is the scalar [7]

$$|z| = \sqrt{z^*z}.$$

A quaternion is called a unit quaternion if its norm is 1.

5.2 Background

5.2.1 *Fuzzy Gaussian Neural Network (FGNN)*

The four-layer structure of the Fuzzy Gaussian Neural Network (FGNN) is shown in Fig. 3.2 [9, 10].

The Fuzzy Clifford Gaussian Neural Networks [4] are designed to continue the development of neural networks in other than the real domain.

5.2.2 *Using the Clifford Algebras in Neural Computing*

A Clifford Neuron (CN) computes the following function from $C_{p,q}^n$ to $C_{p,q}$:

$$y = \sum_{i=1}^{n} w_i \otimes_{p,q} x_i. \tag{5.13}$$

The Clifford neuron constitutes a generalization of a classical neuron, concerning the set of the real numbers as a Clifford algebra. A Basic Clifford Neuron (BCN) is derived [5, 8] from a Clifford Neuron in the case of $n = 1$ in (5.13) and therefore computes the following function from $C_{p,q}$ to $C_{p,q}$:

$$y = w \otimes_{p,q} x, \tag{5.14}$$

where x represents the neuron input and w being the vector of the weights.

The error because of a training vector is expressed [5] by the formula

$$E_{p,q} = \frac{1}{2}(d - w \otimes_{p,q} x)^2, \tag{5.15}$$

where $\| \cdot \|$ means the standard Euclidean norm, and d is the ideal output of the respective neuron.

The weight association of the $\text{BCN}_{0,1}$ of a complex number, named CBCN (Complex Basic Clifford Neuron) is a dilatation-rotation of 2D Euclidean space.

For a such neuron, the update weight formula is [5]:

$$w(t+1) = w(t) - \eta \cdot \frac{\partial E_{0,1}}{\partial w}, \tag{5.16}$$

where η, $0 < \eta < 1$ is the learning rate.

Taking into account that

$$d - w \otimes_{0,1} x = d_1 - w_1 x_1 + w_2 x_2 + (d_2 - w_1 x_2 - w_2 x_1) e_1$$

one achieves

$$\Delta w_{0,1} = -\frac{\partial E_{0,1}}{\partial w} = -\frac{\partial E_{0,1}}{\partial w_1} - \frac{\partial E_{0,1}}{\partial w_2} e_1,$$

namely

$$\begin{aligned} \Delta w_{0,1} &= (d_1 - w_1 x_1 + w_2 x_2) x_1 - (d_2 - w_1 x_2 - w_2 x_1) x_2 \\ &\quad + [(d_2 - w_1 x_2 - w_2 x_1) x_1 + (d_1 - w_1 x_1 + w_2 x_2) x_2] e_1. \end{aligned}$$

In the case of the Clifford Multilayer Perceptron having CBCN neurons, the inputs are of the form:

$$\{x_1^k + x_2^k e_1\}_{k=1}^{K},$$

K being the number of the patterns from the training lot of the network.

5.3 Fuzzy Clifford Gaussian Neural Network (FCGNN)

We shall discuss in detail how we can use the Clifford algebras in neural computing in order to introduce a Fuzzy Clifford Gaussian Neural Network.

The operations that execute to each layer of our fuzzy neural network will be described.

5.3.1 Basic Equations

The neurons of the first level (*input level*) do not process the signals; they only transmit the information to the next level. The output $O^1_{ki} = x^1_{ki},\ i = \overline{1,n}$ represents the input, n is the number of neurons belonging to the first level of FCGNN (equal to the dimension of the input space) and k is the index of the input vector ($k = \overline{1,K}$). The corresponding equations are

$$g^1_{ki}(x^1_{ki}) = x^1_{ki},\ i = \overline{1,n} \tag{5.17}$$

$$O^1_{ki} = f^1_i(g^1_{ki}) = g^1_{ki}(x^1_{ki}),\ i = \overline{1,n}. \tag{5.18}$$

Each neuron of the *linguistic term layer* (level 2) performs a Gaussian membership function

$$g^1_{kij}(x^2_{ki}, m_{ij}, \sigma_{ij}) = -\left(\frac{(x^2_{ki} - m_{ij})^2}{\sigma^2_{ij}}\right)_{p,q},\ i = \overline{1,n},\ j = \overline{1,K}, \tag{5.19}$$

$$O^2_{kij} = f^2_{ij}(g^2_{kij}) = \exp(g^2_{kij})_{pq} = \exp\left(-\frac{(x^2_{ki} - m_{ij})^2}{\sigma^2_{ij}}\right)_{p,q}, \tag{5.20}$$

where the corresponding weights to be refined m_{ij} and σ_{ij} denote the mean and variance with respect to A^j_i, ($i = \overline{1,n},\ j = \overline{1,K}$). The number of neurons characterizing this level is nK. Each input x^2_{ki} is transformed by this layer into a fuzzy membership degree.

The third layer (*rule layer*) computes the antecedent matching by the product operation, according to the relations:

$$g^3_{kj}(x^3_{kij}, W^3_{ij}) = \prod_{i=1\ p,q}^{n} W^3_{ij} \otimes_{p,q} x^3_{kij},\ j = \overline{1,K}, \tag{5.21}$$

$$O^3_{kj} = f^3_j(g^3_{kj}) = g^3_{kj}(x^3_{kij}, W^3_{ij}),\ j = \overline{1,K}, \tag{5.22}$$

where W_{ij}^3 is the connection weight between the $(i-1)K+j$-th node of the second level $(i=\overline{1,n})$ and, the j-th neuron of the third level $(j=\overline{1,K})$. We assume $W_{ij}^3=1,\ (\forall)\ i=\overline{1,n},\ j=\overline{1,K}$.

The *output level* (fourth level) performs the defuzzification:

$$g_{kj}^4(x_{ki}^4, W_{ij}^4)=\sum_{i=1}^{K} W_{ij}^4 \otimes_{p,q} x_{ki}^4,\ j=\overline{1,M}. \tag{5.23}$$

We introduce at this level a *Clifford sigmoid activation function* in order to apply the FCGNN for classification:

$$O_{kj}^4=f_j^4(g_{kj}^4)=\left(\frac{1}{1+\exp\left(-\lambda\cdot g_{kj}^4(x_{ki}^4, W_{ij}^4)\right)}\right)_{p,q},\ j=\overline{1,M},\ \lambda\in\Re \tag{5.24}$$

where W_{ij}^4 is the connection between the neuron i $(i=\overline{1,K})$ of the third level and the neuron j $(j=\overline{1,M})$ of the fourth level.

5.3.2 On-Line Weight Initialization

The mean initialization algorithm consists in:

$$\begin{cases} m_{ik}=x_{ki},\ 1\le k\le K-2,\ 1\le i\le n \\ m_{ik}=a_i,\ k=K-1,\ 1\le i\le n \\ m_{ik}=b_i,\ k=K-2,\ 1\le i\le n \end{cases} \tag{5.25}$$

where

$$a_i=\min_{k=1}^{K} x_{ki} \tag{5.26}$$

and

$$b_i=\max_{k=1}^{K} x_{ki}. \tag{5.27}$$

The variance initialization algorithm is:

$$\sigma_{ik}=\begin{cases} \frac{m_{ik}-m_{iR}}{\sqrt{\ln\alpha_i}},\ \text{if}\ \parallel m_{ik}-m_{iR}\parallel\ge\parallel m_{ik}-m_{iL}\parallel \\ \frac{m_{ik}-m_{iL}}{\sqrt{\ln\alpha_i}},\ \text{if}\ \parallel m_{ik}-m_{iR}\parallel\le\parallel m_{ik}-m_{iL}\parallel, \end{cases} \tag{5.28}$$

$i = \overline{1, n},\ k = \overline{1, K}$, where:

- $\alpha_i, i = \overline{1, n}, 0 <\| \alpha_i \|< 1$ are overlapping factors,
- A_i^R is the closest fuzzy set of A_i^k, on the right of A_i^k,
- A_i^L is the closest fuzzy set of A_i^k, on the left side of A_i^k,
- m_{iR} is the center of A_i^R,
- m_{iL} is the center of A_i^L.

By choosing the means $m_{ik},\ i = \overline{1, n},\ k = \overline{1, K}$ using (5.25) and the variances $\sigma_{ik},\ i = \overline{1, n},\ k = \overline{1, K}$ based on (5.28), the membership functions of the linguistic labels $A_i^j,\ i = \overline{1, n},\ k = \overline{1, K}$ satisfy the Theorem 5.1.

Theorem 5.1 *Let the fuzzy set* $A_i = (A_i^1, A_i^2, \ldots, A_i^k)$, *in which each linguistic label* $A_i^j,\ i = \overline{1, n},\ k = \overline{1, K}$ *has associated a gaussian membership function built on the on-line weight initialization (with the means* $m_{ik},\ i = \overline{1, n},\ k = \overline{1, K}$ *given by (5.25) and the variances* $\sigma_{ik},\ i = \overline{1, n},\ k = \overline{1, K}$ *expressed in (5.28)).*

In this hypotheses, for all $x_i \in X,\ i = \overline{1, n}$ *(in the theorem* x_i *constitutes the arguments of the membership functions, resulted from the respective input vector and* X *represents the set of the all these arguments) there is* $k = \overline{1, K}$ *such that* $\| \mu_{A_i^k} \|\geq\| \alpha_i \|$, *where* $\alpha_i,\ i = \overline{1, n},\ 0 <\| \alpha_i \| < 1$ *are the overlapping factors.*

Proof As $x_i \in X,\ i = \overline{1, m}$ it must exist $k = \overline{1, K}$, such that: $\| m_{ik} \|\leq\| x_i \|\leq\| m_{iR} \|$ or $\| m_{iL} \|\leq\| x_i \|\leq\| m_{ik} \|$.

We shall proof the Theorem 5.1, in the following four cases:

1. $\| m_{ik} \|\leq\| x_i \|\leq\| m_{iR} \|$ and $\| m_{ik} - m_{iR} \|\geq\| m_{ik} - m_{iL} \|$;
2. $\| m_{iL} \|\leq\| x_i \|\leq\| m_{ik} \|$ and $\| m_{ik} - m_{iR} \|\geq\| m_{ik} - m_{iL} \|$;
3. $\| m_{ik} \|\leq\| x_i \|\leq\| m_{iR} \|$ and $\| m_{ik} - m_{iR} \|\leq\| m_{ik} - m_{iL} \|$;
4. $\| m_{iL} \|\leq\| x_i \|\leq\| m_{ik} \|$ and $\| m_{ik} - m_{iR} \|\leq\| m_{ik} - m_{iL} \|$.

Case 1. If $\| m_{ik} \|\leq\| x_i \|\leq\| m_{iR} \|$ and $\| m_{ik} - m_{iR} \|\geq\| m_{ik} - m_{iL} \|$ it results that

$$\sigma_{ik} = \frac{m_{ik} - m_{iR}}{\sqrt{\ln \alpha_i}}.$$

We shall have

$$\| m_{ik} \|\leq\| x_i \|\leq\| m_{iR} \|\Longrightarrow \| x_i - m_{ik} \|\leq\| m_{iR} - m_{ik} \|\Longrightarrow$$

$$\left\| \exp\left(-\frac{(x_i - m_{ik})^2}{\sigma_{ik}^2}\right)_{p,q} \right\| \geq \left\| \exp\left(-\frac{(m_{iR} - m_{ik})^2}{\sigma_{ik}^2}\right)_{p,q} \right\| =\| \exp(-\ln \alpha_i)_{p,q} \|,$$

namely

$$\| \mu_{A_i^k}(x_{ik}) \|\geq\| (\alpha_i)_{p,q}^{-1} \|\geq\| \alpha_i \|,$$

where $\alpha_i,\ i = \overline{1, n},\ 0 <\| \alpha_i \|< 1$ are the overlapping factors.

Case 2. If $\| m_{iL} \| \leq \| x_i \| \leq \| m_{ik} \|$ and $\| m_{ik} - m_{iR} \| \geq \| m_{ik} - m_{iL} \|$ it results that

$$\sigma_{ik} = \frac{m_{ik} - m_{iR}}{\sqrt{\ln \alpha_i}}.$$

We shall have

$$\| m_{iL} \| \leq \| x_i \| \leq \| m_{ik} \| \Longrightarrow \| m_{ik} - m_{iR} \| \geq \| m_{ik} - m_{iL} \| \geq \| m_{ik} - x_i \| \Longrightarrow$$

$$\left\| \exp\left(-\frac{(x_i - m_{ik})^2}{\sigma_{ik}^2}\right)_{p,q} \right\| \geq \left\| \exp\left(-\frac{(m_{iR} - m_{ik})^2}{\sigma_{ik}^2}\right)_{p,q} \right\| = \| \exp(-\ln \alpha_i)_{p,q} \|,$$

namely

$$\| \mu_{A_i^k}(x_{ik}) \| \geq \| \mu_{A_i^k}(x_{ik}) \| \geq \| (\alpha_i)^{-1}{}_{p,q} \| \geq \| \alpha_i \| .$$

Case 3. If $\| m_{ik} \| \leq \| x_i \| \leq \| m_{iR} \|$ and $\| m_{ik} - m_{iR} \| \leq \| m_{ik} - m_{iL} \|$ it results that

$$\sigma_{ik} = \frac{m_{ik} - m_{iL}}{\sqrt{\ln \alpha_i}}.$$

We shall have:

$$\| m_{ik} \| \leq \| x_i \| \leq \| m_{iR} \| \Longrightarrow \| x_i - m_{ik} \| \leq \| m_{iR} - m_{ik} \| \leq \| m_{ik} - m_{iL} \| \Longrightarrow$$

$$\left\| \exp\left(-\frac{(x_i - m_{ik})^2}{\sigma_{ik}^2}\right)_{p,q} \right\| \geq \left\| \exp\left(-\frac{(m_{ik} - m_{iL})^2}{\sigma_{ik}^2}\right)_{p,q} \right\| = \| \exp(-\ln \alpha_i)_{p,q} \|,$$

namely

$$\| \mu_{A_i^k}(x_{ik}) \| \geq \| (\alpha_i)^{-1}{}_{p,q} \| \geq \| \alpha_i \| .$$

Case 4. If $\| m_{iL} \| \leq \| x_i \| \leq \| m_{ik} \|$ and $\| m_{ik} - m_{iR} \| \leq \| m_{ik} - m_{iL} \|$ it results that

$$\sigma_{ik} = \frac{m_{ik} - m_{iL}}{\sqrt{\ln \alpha_i}}.$$

We shall have:

$$\| m_{iL} \| \leq \| x_i \| \leq \| m_{ik} \| \Longrightarrow \| m_{ik} - m_{iL} \| \geq \| m_{ik} - x_i \| \Longrightarrow$$

$$\left\| \exp\left(-\frac{(x_i - m_{ik})^2}{\sigma_{ik}^2}\right)_{p,q} \right\| \geq \left\| \exp\left(-\frac{(m_{iL} - m_{ik})^2}{\sigma_{ik}^2}\right)_{p,q} \right\| = \| \exp(-\ln \alpha_i)_{p,q} \|,$$

namely

$$\| \mu_{A_i^k}(x_{ik}) \| \geq \| (\alpha_i)^{-1}{}_{p,q} \| \geq \| \alpha_i \| .$$

■

5.3.3 Training Algorithm

The training algorithm is of type backpropagation (BP), in order to minimize the error function

$$E = \frac{1}{K} \cdot \sum_{k=1}^{K} E_k, \tag{5.29}$$

where

$$(E_k)_{p,q} = \frac{1}{2} \cdot \sum_{i=1}^{M} (d_{ki} - y_{ki})^2_{p,q}, \; k = \overline{1, K}, \tag{5.30}$$

represents the error for the rule k.

The weight refinement is given by the gradient rule

$$\theta_{ij}(t+1) = \theta_{ij}(t) - \eta \cdot \frac{\partial (E_k)_{p,q}}{\partial \theta_{ij}}, \tag{5.31}$$

where:

- θ_{ij} is the parameter to be refined,
- t is the iteration index,
- η is the learning rate $(0 < \eta < 1)$,
- $d_k = (d_{k1}, \ldots, d_{kM})^t$ is the ideal output vector of the FCGNN when at its input is applied the vector having the index k,
- $y_k = (y_{k1}, \ldots, y_{kM})^t$ is the corresponding real output vector of the FCGNN ($k = \overline{1, K}$).

The training of this neural network is supervised, namely for the K vectors from the training lot we know the set of the ideal outputs.

The refining of the FCGNN parameters can be divided into two phases, depending on the parameters of premises and respective of conclusions of the rules, as follows:

(a) in the part of the premise of the rules, the means and variances of the Gaussian functions one refine.
(b) in the conclusions of the rules, the weights relating to the latest layer of FCGNN must to be refined, the others being equal to 1.

The training algorithm steps are

Step 1 Apply the vector $X_k = (x_{k1}, \ldots, x_{kn})^t$, corresponding to the rule $\Re_k$, to the network input. Initialize the weights W_{ij}^4 related to output layer (the only network weights that are not equal to 1, as the weights corresponding to the third layer are equal to 1) with some random values, uniformly distributed in the interval $[-0.5, 0.5]$. Initialize the means m_{ij} and the variances $\sigma_{ij},\ i = \overline{1, n},\ j = \overline{1, K}$, according to the on-line initialization algorithm.

Step 2 Compute the neuron aggregations in the first layer, using the relation (5.17).

Step 3 Calculate outputs of the neurons from the first layer of FCGNN, based on the relation (5.18).

Step 4 Calculate the inputs of neurons in the second layer, using the relation (5.19).

Step 5 Compute the neuron activations in the second layer, using the relation (5.20).

Step 6 Use the relation (5.21) to determine the inputs of the neurons from the third layer.

Step 7 Use the relation (5.22) in order to compute the outputs of the neurons from the third layer.

Step 8 Calculate the neuron aggregations in the output layer, using the relation (5.23).

Step 9 Compute the outputs of the neurons from the fourth layer with (5.24).

Step 10 Calculate error terms for the neurons from the output layer, on the basis of relation:

$$\delta^4_{pj} = \lambda \cdot (d^4_{pj} - O^4_{pj}) \otimes_{p,q} O^4_{pj} \otimes_{p,q} (1 - O^4_{pj}),\ j = \overline{1, M}. \tag{5.32}$$

Step 11 Refine the weights corresponding to the output layer, according to the formula:

$$W^4_{ij}(t+1) = W^4_{ij}(t) + \eta_W \cdot \delta^4_{pj} \otimes_{p,q} x^4_{pi},\ i = \overline{1, K},\ j = \overline{1, M}, \tag{5.33}$$

η_W being a learning rate for the weights.

Step 12 Compute the following error terms:

$$\delta^3_{pj} = -\sum_{i=1}^{M} \delta^4_{pi} \otimes_{p,q} W^4_{ji},\ j = \overline{1, K}. \tag{5.34}$$

Step 13 Calculate the following error terms:

$$\delta^2_{pj} = \delta^3_{pj} \otimes_{p,q} \prod_{i=1\ p,q}^{n} O^2_{pij},\ j = \overline{1, K}. \tag{5.35}$$

Step 14 Refine the means corresponding to the membership functions O^2_{pij}:

$$m_{ij}(t+1) = m_{ij}(t) + \eta_m \cdot \delta^2_{pj} \otimes_{p,q} 2\frac{x^2_{ki} - m_{ij}}{\sigma^2_{ij}}, \; i = \overline{1,n}, \; j = \overline{1,K}, \tag{5.36}$$

η_m being a learning rate for the means.

Step 15 Refine the means corresponding to the membership functions O^2_{pij}:

$$\sigma_{ij}(t+1) = \sigma_{ij}(t) + \eta_\sigma \cdot \delta^2_{pj} \otimes_{p,q} 2\frac{(x^2_{ki} - m_{ij})^2}{\sigma^3_{ij}}, \; i = \overline{1,n}, \; j = \overline{1,K}, \tag{5.37}$$

η_σ being a learning rate for the variances.

Step 16 Find the error because of the training vector k through the relation (5.30).

Step 17 If $k < K$ (i.e. it isn't traversed the entire training lot) one proceed to the next vector from the training lot and one repeat the algorithm from the *Step 2*.

Step 18 Compute the error corresponding to the respective epoch of the training algorithm, according to the formula (5.29).

Step 19 Test the stop condition of the training algorithm, which is after a fixed number of epochs. If the condition is accomplished, the algorithm stops. Otherwise, we begin a new epoch of learning.

5.4 Experimental Evaluation

5.4.1 2D Rotation with Euler's Equation

The unit circle means an important part in the theory of complex numbers, every point on the circle being the form

$$z = \cos\theta + \mathrm{i}\sin\theta = \mathrm{e}^{\mathrm{i}\theta}. \tag{5.38}$$

Euler's formula (5.38) can be used [7] to represent a 2D point with a length and angle on a complex plane; raising the constant e to the imaginary power $\mathrm{i}\theta$ creates the complex number with the angle θ in radians.

"This polar form of complex exponential function, $\mathrm{e}^{\mathrm{i}\theta}$ is very convenient to represent rotating objects or periodic signals because it can represent a point in the complex plane with only single term instead of two terms, $x + \mathrm{i}y$."[6]; it simplifies the mathematics when used in multiplication:

$$\mathrm{e}^x \cdot \mathrm{e}^{\mathrm{i}y} = \mathrm{e}^{x+\mathrm{i}y}.$$

[6] Ahn, S. H, 2009. Math. http://www.songho.ca/math/index.html.

In 2D, the multiplication of two complex numbers implies 2D rotation, about the origin. The sum of angles in the imaginary exponential form equals to the product of two complex numbers,

$$e^{i(\theta+\phi)} = e^{i\theta} \cdot e^{i\phi}.$$

It tells us multiplying $e^{i\theta}$ by $e^{i\phi}$ performs rotating $e^{i\theta}$ with angle ϕ. When $z = x+iy$ is multiplied by $e^{i\phi}$, the length of $z' = z \cdot e^{i\phi}$ remains the same ($|z| = |z'|$), but the angle of z' is added by ϕ.

5.4.2 3D Rotation with Quaternion

Hamilton was inspired by the solution to the analogous problem in two dimensions: rotations of the plane about the origin can be encoded by unit length complex numbers.

Just as the unit circle is important for planar polar coordinates, so the 3-sphere is important in the polar view of 4-space involved in quaternion multiplication.

However, multiplying a quaternion p by a unit quaternion z doesn't conserve the length (norm) of the vector part of the quaternion p [7]. Thus, we need a special multiplication for 3D rotations that is length-conserving transformation. For this reason, we multiply the unit quaternion z at the front of p and multiplying z^* (called the conjugation by z) at the back of p, in order to cancel out the length changes.

A rotation about the origin in $\Re^3$ can be specified by giving a vector for the axis of rotation and an angle of rotation about the axis. The vector $v = (x\ y\ z) \in \Re^3$ can be represented by the quaternion $z = 0 + xi + yj + zk$.

The representation of a rotation as a quaternion (4 numbers) is more compact than the representation as an orthogonal matrix (9 numbers). We will use the following four real numbers to specify a rotation: three coordinates for a vector and one real number to give the angle. This is far fewer than the nine entries of a 3×3 orthogonal matrix we know to use in Linear Algebra.

Let us consider a unit quaternion $z = q0 + q$ only. That $q_0^2 + |q|^2$ implies that there must exist [6] some angles $\theta \in [0, \pi]$ such that

$$\begin{cases} \cos^2\theta = q_0^2 \\ \sin^2\theta = |q|^2. \end{cases}$$

In fact, there exists a unique $\theta \in [0, \pi]$ such that

$$\begin{cases} \cos\theta = q_0 \\ \sin\theta = |q|. \end{cases}$$

The unit quaternion can now be written in terms of the angle θ and the unit vector $u = q/|q|$:

$$z = \cos\theta + u \sin\theta, \tag{5.39}$$

where u as a 3D vector has length 1.

We can note the similarity to unit length complex numbers $\cos\theta + \mathrm{i}\sin\theta$. In fact, Euler's identity for complex numbers generalizes to quaternions:

$$e^{u\theta} = \cos\theta + u\sin\theta. \tag{5.40}$$

Using the unit quaternion z we define [6] an operator on vectors $v \in \Re^3$:

$$L_z(v) = zvz^* = (q_0^2 - |q|^2)v + 2(q \cdot v)q + 2q_0(q \times v), \tag{5.41}$$

where $q \cdot v$ means the inner product and $q \times v$ is the cross product of the vectors v and q from $\Re^3$.

One can notice that

1. the quaternion operator (5.41) does not change the length of the vector v, namely $|L_z(v)| = |v|$;
2. the direction of v, if along z, is left unchanged by the operator $L_z(v)$.

The two previous observations make us guess the operators L_z acts like a rotation about q.

Theorem 5.2 ([6]) *For any unit quaternion*

$$z = q_0 + q = \cos\frac{\theta}{2} + u\sin\frac{\theta}{2}, \tag{5.42}$$

and for any vector $v \in \Re^3$, the action of the operator $L_z(v) = zvz^*$ on v is equivalent to a rotation of the vector through an angle θ about q as the axis of rotation.

5.4.3 Data Sets

Using MATLAB 7.9, we shall generate ten classes of 2D geometrical shapes (ten patterns for each of class, see Fig. 5.1), like: Ellipses, Cardioides, Astroides (hypocycloid with four cusps), Parabolas, Lemniscates of Bernoulli, Cycloides, Nefroides, Butterfly curves, Tricuspoids, Conchoids of de Sluze.

We need the following sequence in MATLAB 7.9 in order to build an astroide:

```
>> a = 2;
>> t = 0 : 2 * pi/99 : 2 * pi;
>> x = 3 * a * cos(t) + a * cos(3 * t);
>> y = 3 * a * sin(t) - a * sin(3 * t);
>> plot(x, y)
```

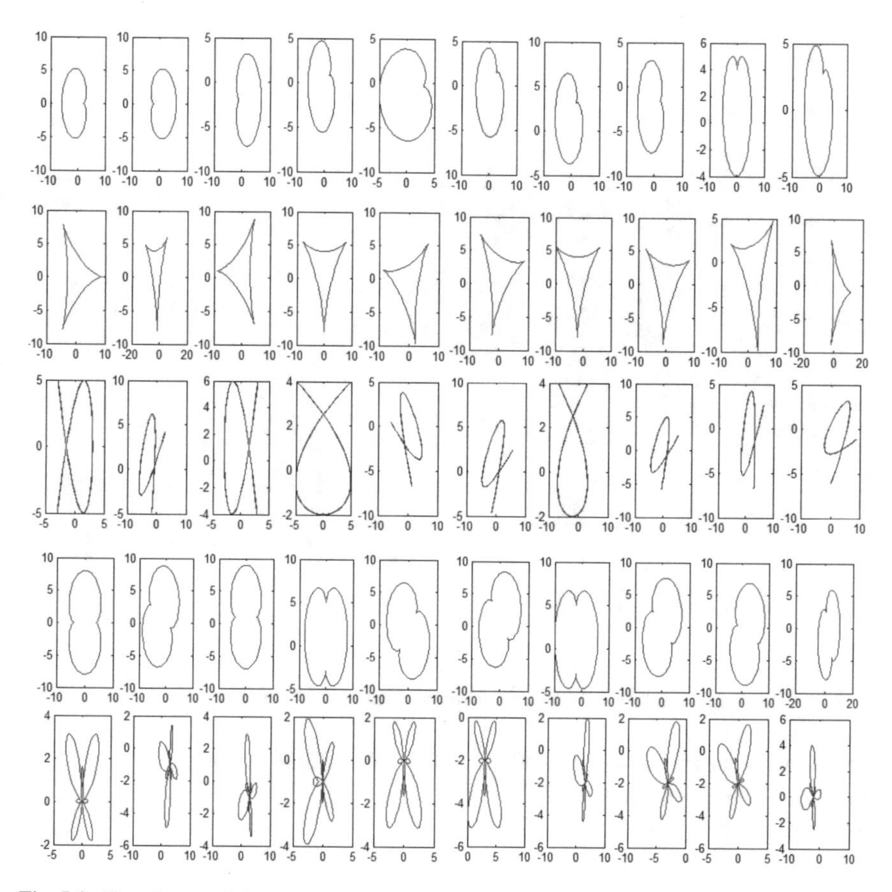

Fig. 5.1 Five classes of plane curves

and the following MATLAB 7.9 code to generate an astroide resulted by the rotation of the first astroide with the angle $(pi/9)$, around the origin and by the translation with the vector $0 - 2i$:

$>> z1 = (x + i * y) * (cos(pi/9) + i * sin(pi/9)) - 2 * i;$

$>> x1 = real(z1);$

$>> y1 = imag(z1);$

$>> plot(x1, y1)$

The half from these patterns constitute the training lot and the others comprise the test lot. We shall use a Fuzzy Clifford Gaussian Neural with Complex Basic Clifford Neurons to recognize the 2D shapes.

Similarly, we shall generate through MATLAB 7.9, other ten classes of 3D geometrical shapes (ten patterns for each of class, (see Fig. 5.2): Hyperboloids of one sheet, Cylinders, Hyperbolic paraboloids, Elliptic paraboloids, Rotating cones, Bohemian domes, Astroidal Ellipsoids, Helices, Toruses.

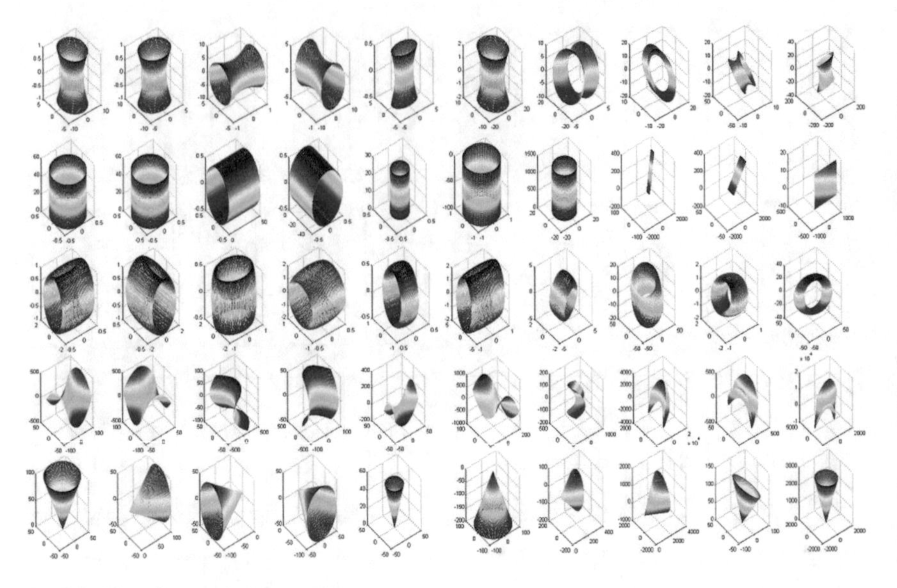

Fig. 5.2 Five classes of plane curves

The following Matlab 7.9 code generates an elliptic paraboloid

```
>> a = 3; b = 2;
>> [x, y] = meshgrid(−60 : 150/99 : 90, −50 : 100/99 : 50);
>> z = x.^2/(2 * a^2) + y.^2/(2 * b^2);
>> mesh(x, y, z)
```

while the following code:

```
>> q = [−sqrt(2)/2 0 sqrt(2)/2 0];
>> for i = 1 : 100
for j = 1 : 100
w = [0 x(i, j) y(i, j) z(i, j)];
u = quatmultiply(q, w);
r = quatmultiply(u, quatconj(q));
x2(i, j) = r(2);
y2(i, j) = r(3);
z2(i, j) = r(4);
end
end
>> mesh(x2, y2, z2)
```

is equivalent to a rotation of the elliptic paraboloid through an angle $(3\pi/2)$, around the origin, about [0 1 0] as the axis of rotation.

Table 5.1 The recognition rate over the test lot, using FCGNN, CMP.

	Performance on	
	FGNN (%)	CMP (%)
2D curve recognition	98	96
3D surface recognition	96	94

Our proposed Clifford neural network needs quaternionic neurons in order to be used to the recognition of the 3D shapes.

5.4.4 Experimental Results

We have experimented the recognition task of some 2D and respective 3D geometrical shapes for a database of 100 images from 10 classes (where 50 images are in the training lot and the other in the test lot). The training lot consists of that images which are already labeled.

We have performed the software implementation of our model using Matlab 7.9.

We shall compare the performances of the FCGNN with those obtained using a CMP (Clifford Multilayer Perceptron, described in [5]). The experimental results are shown in Table 5.1.

The similarities between FCGNN and CMP are:

1. the both networks have a training backpropagation algorithm;
2. the activation functions of each layer are nonlinear functions;
3. both FCGNN learning and CMP learning is supervised;
4. each of the neurons of the two networks performs two operations, using two different functions: the aggregation function and the activation function;
5. the input layer of the both networks is a transparent layer.

The differences between FCGNN and CMP are:

1. FCGNN has many intermediate layers, while CMP only two intermediate layers;
2. FCGNN has a special connection between the second and third layer, in comparison with CMP, which has a normal connection between these layers;
3. FCGNN has some fuzzy outputs for the second and third layer, while CMP has only non-fuzzy outputs.

The advantage of FCGNN is that for some certain values of the overlapping parameters we obtain the best recognition rates on the test lot, compared with the CMP. The disadvantage of FCGNN is that it requires a large number of term neurons (on the second layer), namely $n \times K$ neurons for n inputs, and K fuzzy rules.

References

1. B.W. York and Q. Yi. Clifford neural networks and chaotic time series. http://www.deepstem.com/Bio/SysSci-CNN-york.pdf, 2013.
2. E. Bayro-Corrochano, R. Vallejo, and N. Arana-Daniel. Geometric preprocessing, geometric feedforward neural networks and clifford support vector machines for visual learning. *Neurocomputing*, 67:54–105, 2005.
3. L. Ding, H. Feng, and L. Hua. Multi-degree of freedom neurons geometric computation based on Clifford algebra. In *Fourth International Conference on Intelligent Computation Technology and Automation*, pages 177–180, 2011.
4. I. Iatan and M. de Rijke. A fuzzy Gaussian Clifford neural network. (work in progress), 2014.
5. S. Buchholz and G. Sommer. On Clifford neurons and Clifford multi-layer perceptrons. *Neural Networks*, 21:925–935, 2008.
6. Y.B. Jia. Quaternions and rotations. http://www.cs.iastate.edu/~cs577/handouts/quaternion.pdf, 2013.
7. S. H. Ahn. Math. http://www.songho.ca/math/index.html, 2009.
8. S. Buchholz and G. Sommer. Introduction to neural computation in Clifford algebra. http://www.informatik.uni-kiel.de/inf/Sommer/doc/Publications/geocom/buchholz_sommer1.pdf, 2010.
9. V. Neagoe, I. Iatan, and S. Grunwald. A neuro-fuzzy approach to ecg signal classification for ischemic heart disease diagnosis. In *the American Medical Informatics Association Symposium (AMIA 2003), Nov. 8–12 2003, Washington DC*, pages 494–498, 2003.
10. V. Neagoe and I. Iatan. Face recognition using a fuzzy-gaussian neural network. In *Proceedings of First IEEE International Conference on Cognitive Informatics, ICCI 2002, 19–20 August 2002, Calgary, Alberta, Canada*, pages 361–368, 2002.

Chapter 6
Concurrent Fuzzy Neural Networks

The aim of this chapter is to introduce two concurrent fuzzy neural network approaches for a:

(1) Fuzzy Nonlinear Perceptron (FNP) and
(2) Fuzzy Gaussian Neural Network (FGNN),

each of them representing a winner-takes-all collection of fuzzy modules.

Our two proposed models, entitled:

(1) *Concurrent Fuzzy Nonlinear Perceptron Modules* (CFNPM)
(2) *Concurrent Fuzzy Gaussian Neural Network Modules* (CFGNNM)

will be applied for the pattern classification.

For the task of recognition, we have used a processing cascade having two stages:

(a) Feature extraction using either the Principal Component Analysis (PCA) or the Discrete Cosine Transform (DCT)
(b) Pattern classification using FNP, CFNPM, FGNN, and CFGNNM.

Feature extraction is the most important stage of the recognition task as the pattern classification is completely dependent on this stage.

6.1 Baselines

6.1.1 Principal Component Analysis

The Karhunen–Loève Transformation (KLT) or Principal Component Analysis (PCA) is [1, 2] an orthogonal transformation, which allows the optimal representation of a vector X for the mean square error criterion.

I.F. Iatan, *Issues in the Use of Neural Networks in Information Retrieval*,
Studies in Computational Intelligence 661, DOI 10.1007/978-3-319-43871-9_6

The transformation matrix is

$$\mathrm{K} = (\Phi_1, \ldots, \Phi_n)^t,$$

where Φ_i are n-dimensional vectors that need to be determined, having the property:

$$\Phi_i^t \cdot \Phi_j = \begin{cases} 1, & \text{if } i = j \\ 0, & \text{if } i \neq j. \end{cases} \tag{6.1}$$

Therefore,

$$\mathrm{K}^t \cdot \mathrm{K} = \mathrm{K} \cdot \mathrm{K}^t = \mathrm{I}_n, \tag{6.2}$$

I_n is the *identity matrix* or *unit matrix* of size n.

The vector X_p, $p = \overline{1, N}$ (N being the number of the vectors that have to be transformed) will be transformed into the vector

$$Y_p = \mathrm{K} \cdot X_p = (y_{p1}, \ldots, y_{pn})^t, \tag{6.3}$$

with

$$y_{pi} = \Phi_i^t \cdot X_p, \ (\forall)\ i = \overline{1, n}. \tag{6.4}$$

The transformation defined by (6.3) is called the *Karhunen–Loève Transformation*.

Using the relations (6.2) and (6.3) it will result:

$$X_p = \mathrm{K}^t \cdot Y_p = \sum_{i=1}^{n} y_{pi} \cdot \Phi_i. \tag{6.5}$$

The formula (6.5) represents the *Karhunen–Loève Transformation expansion*.

We shall only keep $m < n$ component of the vector Y_p, the other $n - m$ components being replaced with the preselected constant b_i, such that one estimates:

$$\hat{X}_p = \sum_{i=1}^{m} y_{pi} \cdot \Phi_i + \sum_{i=m+1}^{n} b_i \cdot \Phi_i.$$

The error corresponding to this estimation is

$$\Delta X_p(m) = X_p - \hat{X}_p(m) = \sum_{i=m+1}^{n} (y_{pi} - b_i) \cdot \Phi_i.$$

We shall assume the mean square error criterion:

$$\overline{\varepsilon}_p^2(m) = \mathrm{E}\{\|\Delta X_p(m)\|^2\} = \mathrm{E}\left\{\sum_{i=m+1}^{n}\sum_{j=m+1}^{n}(y_{pi}-b_i)(y_{pj}-b_j)\Phi_i^t\Phi_j^t\right\}; \tag{6.6}$$

taking into account the relation (6.1), the formula (6.6) becomes

$$\overline{\varepsilon}_p^2(m) = \sum_{i=m+1}^{n}\mathrm{E}\{(y_{pi}-b_i)^2\}, \tag{6.7}$$

E meaning the expectation operator.

The error function $\overline{\varepsilon}_p^2(m)$ can be minimized by choosing Φ_i and b_i adequately; this process is performed in two stages:

Stage 1.

Minimize $\overline{\varepsilon}_p^2(m)$ relative with b_i, putting the condition that

$$\frac{\partial}{\partial b_i}\mathrm{E}\{(y_{pi}-b_i)^2\} = -2\cdot\mathrm{E}\{(y_{pi}-b_i)\} = 0. \tag{6.8}$$

Basis on the relation (6.4), from (6.8) we achieve

$$b_i = \mathrm{E}\{y_{pi}\} = \Phi_i^t\cdot\mathrm{E}\{X_p\},\ (\forall)\ i=\overline{1,n}. \tag{6.9}$$

Then, the mean square error becomes

$$\overline{\varepsilon}_p^2(m) = \sum_{i=m+1}^{n}\Phi_i^t\Sigma_{X_p}\Phi_i, \tag{6.10}$$

where Σ_{X_p} constitutes the covariance matrix of the vector X_p, namely

$$\Sigma_{X_p} = \mathrm{E}\{\left(X_p-\mathrm{E}\{X_p\}\right)\cdot\left(X_p-\mathrm{E}\{X_p\}\right)^t\}. \tag{6.11}$$

Stage 2.

Minimize $\overline{\varepsilon}_p^2(m)$ relative with Φ_i to find the optimal Φ_i, by imposing the conditions (6.1).

We shall use the method of Lagrange multipliers, which involves the minimization of the expression

$$e_p(m) = \overline{\varepsilon}_p^2(m) - \sum_{i=m+1}^{n}\beta_i(\Phi_i^t\Phi_i-1), \tag{6.12}$$

which is equivalently (based on (6.10)) with

$$e_p(m) = \sum_{i=m+1}^{n} [\Phi_i^t \Sigma_{X_p} \Phi_i - \beta_i(\Phi_i^t \Phi_i - 1)]; \tag{6.13}$$

the expression has to be minimized relative with Φ_i, where β_i are the Lagrange multipliers.

In this aim, we shall impose the condition

$$\nabla_{\Phi_i} e_p(m) = 0; \tag{6.14}$$

taking into account that

$$\nabla_{\Phi_i}(\Phi_i^t \Sigma_{X_p} \Phi_i) = 2\Sigma_{X_p} \Phi_i \tag{6.15}$$

and

$$\nabla_{\Phi_i}(\Phi_i^t \Phi_i) = 2\Phi_i, \tag{6.16}$$

the formula (6.14) becomes

$$\Sigma_{X_p} \Phi_i = 2\beta_i \Phi_i. \tag{6.17}$$

Equation (6.17) proves that Φ_i is an eigenvector of the covariance matrix Σ_{X_p} and β_i is its corresponding eigenvalue, which will be denoted with λ_i $(\forall)$ $i = \overline{1, n}$.

Introducing (6.17) into (6.13), it results:

$$[\overline{\varepsilon}_p^2(m)]_{\min} = \sum_{i=m+1}^{n} \lambda_i. \tag{6.18}$$

The importance of each selected feature is determined by the corresponding eigenvalue. If the eigenvalues are indexed such that: $\lambda_1 > \cdots > \lambda_n > 0$ (the eigenvalues are real and positive as the matrix Σ_{X_p} is symmetric), then the retention priority of the features y_{pi} is in the natural order of the indices.

6.2 Face Recognition Using the Stage of the Feature Selection with PCA/DCT

The aim of applying the two transformations of feature selection:

(1) Principal Component Analysis (PCA) or Karhunen–Loève Transformation (KLT),
(2) Discrete Cosine Transformation (DCT)

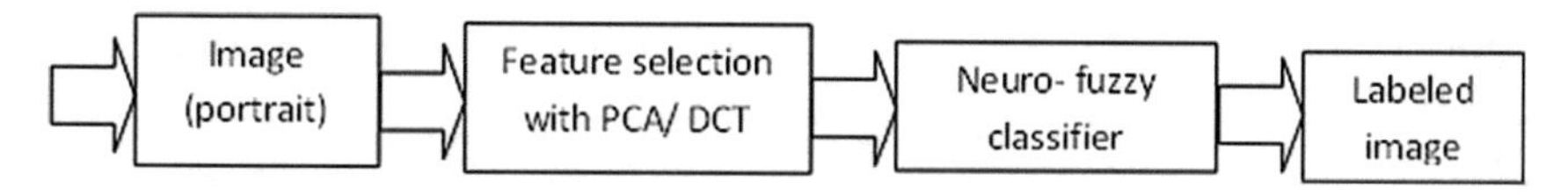

Fig. 6.1 The system for face recognition

is to reduce the dimension of the input data through the selection of those features that contain the most information (the most energy is concentrated inside them) about the considered vector.

Figure 6.1 shows that a face recognition system consists [3] in the following two components:

(a) the feature selection, using PCA/DCT;
(b) the neuro-fuzzy classifier.

The original images belonging to the ORL Database of Faces, with the size of 92 × 112 pixels have been reduced to the dimension of 46 × 56 in order to perform the stage of the feature selection with PCA/DCT; the changing of the image size is imposed by the complex computations that occur in the realization of the PCA/DCT.

We shall analyze [3] the following two situations:

- 10 classes of subjects and 100 images (each of the 10 subjects appears in 10 instances). Based on these images, the vectors X_i, $i = \overline{1, 100}$, with 2576 components ($24 \times 46 = 2576$) have resulted;
- 40 classes of subjects and 400 images (each of the 40 subjects appears in 40 instances). From these images, the vectors X_i, $i = \overline{1, 400}$, having 2576 components were formed.

In the case of applying the feature selection with PCA, the covariance matrix has the dimension 2576 × 2576 and it has been computed using:

(a) 50 images of 10 classes;
(b) 200 images of 40 classes,

in both the cases taking five training images from each class.

The matrix K of the KLT has as rows, the 2576 eigen vectors, with $n = 2576$ components, that have been sorted in a descending order.

In the case of DCT, the elements of the transformation matrix T are computed using the formula (1.1).

By applying KLT/DCT, the vectors X_i became the vectors Y_i $(\forall)$ $i = \overline{1, N}$ (see the relations (6.3), respective (1.3)), where N signifies the number of considered vectors, namely: $N = 50$ in the case (a) and $N = 200$ in the case (b).

Table 6.1 The energy conservation factor F [%], as a function of m, the number of the features, selected with PCA/DCT ($M = 10$)

m	PCA/DCT	F [%]
50	**PCA**	**99.99**
	DCT	**97.47**
43	PCA	99.15
	DCT	97.21
39	PCA	98.24
	DCT	97.04
35	PCA	97.1
	DCT	96.83
32	PCA	96.1
	DCT	96.67
30	PCA	95.33
	DCT	96.56
27	PCA	94.02
	DCT	96.37
25	PCA	93.07
	DCT	96.21
23	PCA	92.02
	DCT	96.04
22	PCA	91.48
	DCT	95.94
20	PCA	90.18
	DCT	95.73

From Tables 6.1 and 6.2 is evaluated the energy conservation factor F, which can be computed with the formula:

$$F = \frac{\sum_{i=1}^{m} \lambda_i}{\sum_{i=1}^{n} \lambda_i} \cdot 100 \tag{6.19}$$

in the case of the PCA and respectively:

$$F = \frac{\sum_{p=1}^{N} E(Y_p^2)}{\sum_{p=1}^{n} E(X_p^2)} \cdot 100, \tag{6.20}$$

Table 6.2 The energy conservation factor F [%], as a function of m, the number of the features, selected with PCA/DCT ($M = 40$)

m	PCA/DCT	F [%]
280	**PCA**	**100**
	DCT	**99.01**
250	PCA	100
	DCT	98.92
199	PCA	99.99
	DCT	98.72
190	PCA	99.87
	DCT	98.67
180	PCA	99.66
	DCT	98.62
170	PCA	99.4
	DCT	98.57
160	PCA	99.09
	DCT	98.51
150	PCA	98.72
	DCT	98.44
100	PCA	95.79
	DCT	98.02

for the DCT, where $E(Y_p^2)$ means the mean square of the transformed vector Y_p, being defined in the formula (1.4).

Figure 6.2 illustrates how the energy concentration factor varies, depending on the number of features, selected with PCA/DCT, for $M = 10$ and $M = 40$.

From Tables 6.1 and 6.2 we can notice that, in the both of cases ($M = 10$ and $M = 40$), we need to select a bigger number of features with DCT than PCA to achieve an energy conservation factor of approximately 100 %.

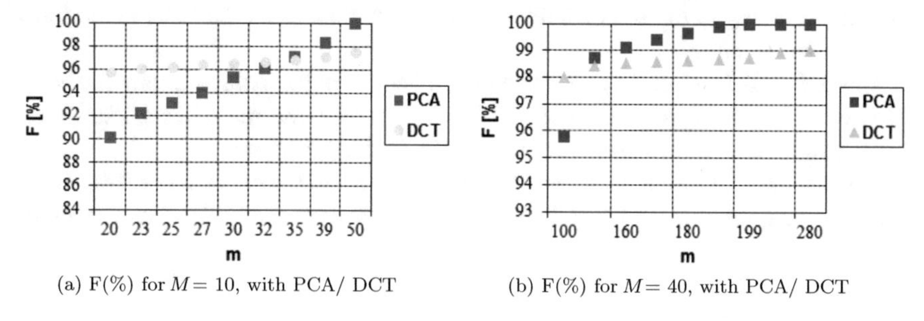

(a) F(%) for M= 10, with PCA/ DCT

(b) F(%) for M= 40, with PCA/ DCT

Fig. 6.2 The energy concentration factor ($M = 10$ and $M = 40$) with PCA/DCT

As bigger as the number of the features selected by applying PCA/DCT, the energy concentration factor will be bigger.

6.3 ECG Classification in the Case of the Feature Selection with PCA/DCT

Definition 6.1 ([1, 3]) The electrocardiogram (EKG or ECG) is a graphical recording of the cardiac phenomena (magnitude and direction) versus time, being used to diagnose various heart disorders.

The *Ischemic (Ischaemic) Heart Disease (IHD)*, otherwise known as *Coronary Artery Disease*, is [4] a condition that affects the supply of the blood to the heart. IHD is the most common cause of death in several countries around the world. Recently, there are many approaches involving techniques for computer processing of 12 lead electrocardiograms, in order to diagnose a certain disease. A first group of methods to interpret the ECG significance uses a morphological analysis. For example, myocardial ischemia may produce a flat or inverted T wave, that is classical narrow and symmetrical. A second group of techniques for computer analysis of ECG uses statistical models.

Figure 6.3[1] shows the ECG which corresponds to a cardiac revolution; it presents more features [5]:

- a wave P corresponding to the depolarization of the atria;
- the interval PR for the conduction time through the tract of Hiss (between the atria and ventricles);
- the complex QRS associated to the ventricular activity;
- the point J which connects the QRS complex by the ST interval;
- the ST interval corresponding to the deactivation phase; it is highest for the women;
- the wave T corresponding to the depolarization of the ventricles.

The peak R is always the point where the signal reaches the maximum amplitude.

The ventricular systolic corresponds to the QT interval. The ventricular diastole corresponds to the interval between the end of the T wave and the beginning of $QRST$ complex. The total diastole corresponds to the area between the end of the wave T and the beginning of the wave P; this phenomenon appears on the ECG in the form of the isoelectric line.

The Ischemic Cardiopathy (IC) occupies a very important place among the heart diseases with a high probability of occurrence. To diagnose the IC was considered that the significant information is concentrated in the $QRST$ area.

We assume that the neuro-fuzzy models used [3] as the classifiers of the ECG signals to diagnose the IC admit the reduced prototypes of the ECG signals as input forms.

---

[1]Neagoe, V. E. and Stănăşilă, O., Pattern Recognition and Neural Networks (in Romanian), 1999, Ed. Matrix Rom, Bucharest.

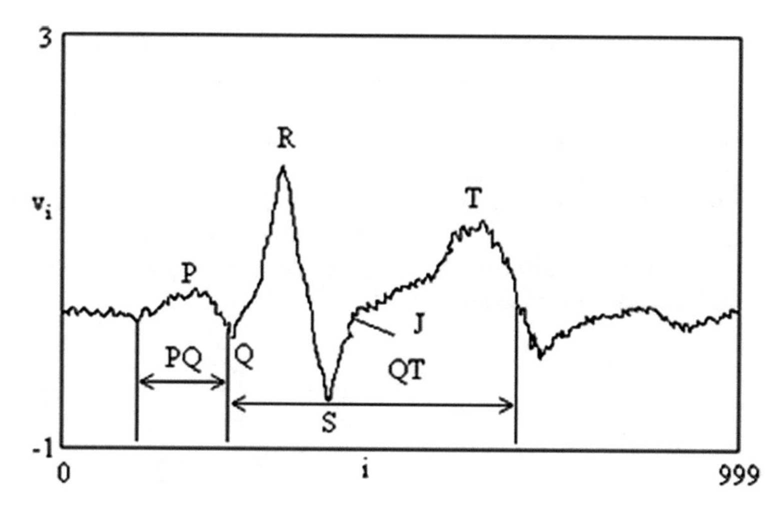

Fig. 6.3 ECG morphology (see footnote 1)

Definition 6.2 ([1, 3]) The *reduced prototype* is the prototype portion corresponding to the *QRST* area.

The reduced prototype have been obtained by processing the ECG signals according to two main stages [1, 3–5]:

1. the statistical analysis of the ECG signals and the generation of the prototype waveform;
2. the compression of the reduced prototype by applying the Principal Component Analysis (PCA) or the Discrete Cosine Transform (DCT).

Definition 6.3 ([1, 3]) The *prototype* constitutes a characteristic waveform, which contains the useful information from a recording.

Initially, the ECG signal is represented by a sequence of samples, having a variable length. From this sequence it must to be selected a characteristic period for the diagnosis. The samples are normalized in amplitude so that the highs to be around 2 mV. This value will be reached only by the peaks *R* and the analysis of the ECG signal may be done according to the following steps [1, 3, 5]:

Step 1 *The detection of the peaks R.* One detects the position of the peaks *R* for each cardiac cycle, using an iterative search technique: one seeks the overcoming of a threshold value ($\approx$2 mV) on the all signal, obtaining some samples compact groups. From these compact groups of samples, only the maximum corresponding the peak *R* will be retained.

Step 2 *The statistical analysis of the cardiac cycle length.* One determines the average length of cycles $\overline{RR}$, excluding from the calculation the first and last maximum, which belongs to some incomplete periods, determining the absence of the peak *R*.

Step 3 *The determination of the maximum area, containing the useful information for each cycle, relative to the position of the peak R.* One considers that a cardiac cycle corresponding to an ECG starts to one-third of the cycle time ($1/3 \cdot \overline{RR}$) before the peak R and finishes to two-thirds of cycle time ($2/3 \cdot \overline{RR}$) after the peak R.

Step 4 *The exclusion the cycles that present high dissimilarities compared to an average cycle and formation of the extended prototype.* One computes the correlation coefficient between a given interval and a baseline interval according to the formula [1, 3]:

$$r = \frac{\sum_{i=1}^{N}(x_i - \overline{x}) \cdot (y_i - \overline{y})}{\sqrt{\left(\sum_{i=1}^{N}(x_i - \overline{x})^2\right) \cdot \left(\sum_{i=1}^{N}(y_i - \overline{y})^2\right)}}, \tag{6.21}$$

where

- $X = (x_1, x_2, \ldots, x_N)$ and $Y = (y_1, y_2, \ldots, y_N)$ are the vectors that characterize the two cardiac cycles whose correlation is evaluated;
- $\overline{x} = \frac{1}{N}\sum_{i=1}^{N} x_i, \overline{y} = \frac{1}{N}\sum_{i=1}^{N} y_i$.

  If r is greater than a critical threshold (=0.96), then the tested interval is excluded, without being considered in the calculation of the prototype. The extended prototype of the cardiac wave can be computed like being the average of the remaining cycles after the exclusion operation.

Step 5 *The location of the area, which contains the significant information (the location of the border points, belonging to the isoelectric line).* We need to retain only the region delimited by the start of the P wave and the end of the T wave from the maximum useful area, determined the *Step 3*. To estimate the positions of the frontier points mentioned above, one makes a piecewise linear approximation, each segment of the approximation corresponding to a group of $(M + 1)$ samples. If $f(i)$ is the digitized ECG sample and the corresponding linear approximation is [5]:

$$z(i) = A \cdot i + B, \ (\forall)\ i = \overline{k, k + M}, \tag{6.22}$$

then the square error of approximation will be [1]:

$$\varepsilon = \sum_{i=k}^{k+M}(A \cdot i + B - f(i))^2. \tag{6.23}$$

The error (6.23) is minimized by imposing the conditions to cancel the partial derivatives of ε with respect to A and B; it results [1]:

$$A = \frac{\sum_{i=k}^{k+M} i \cdot f(i) - \frac{1}{M+1} \cdot \left(\sum_{i=1}^{k+M} i\right) \cdot \left(\sum_{i=k}^{k+M} f(i)\right)}{\sum_{i=k}^{k+M} i^2 - \frac{1}{M+1} \left(\sum_{i=k}^{k+M} i\right)^2} \tag{6.24}$$

$$B = \frac{1}{M+1} \cdot \sum_{i=k}^{k+M} f(i) - \frac{A}{M+1} \cdot \sum_{i=k}^{k+M} i. \tag{6.25}$$

Using the expressions of A and B, given by the relations (6.24) and (6.25) we shall obtain from (6.23) the minimum square error.
By achieving the piecewise linear approximation of several digitized ECG recordings (with the sampling period of 1 ms) for various values of the parameter M, it was found [1, 5] that the optimal value of M is 15.
The location of the isoelectric line corresponds to the area of "bioelectric silence" of the diastole, between the end of the T wave, and the beginning of the wave P.
As the ECG signals contain some neuromuscular noises, that are some perturbations caused by the polarization of the electrodes and other electrical interferences (harmonics of 50 Hz), the location of the isoelectric line is a nontrivial problem.

Definition 6.4 ([1, 3]) The *isoelectric line* is defined like the set of the adjacent areas (each of them containing $M + 1$ samples) that have the slope smaller than a threshold.

To avoid the elimination of the horizontal regional approximations that are located inside the $QRST$ area it is necessary to remove only the first and last such regions of the prototype (focused on the R wave).

Step 6 *Focusing the ECG reduced prototype on the isoelectric line.* The following linear transformation [1, 3, 5], which allows that the signal to be focused on the isoelectric line will be accomplished for every sample:

$$D'(k) = D(k) - \left(D(\alpha_m) + \frac{D(\beta_m) - D(\alpha_m)}{\beta_m - \alpha_m} \cdot (k - \alpha_m)\right), \ (\forall)\ k \in [\alpha_m, \beta_m], \tag{6.26}$$

where

- $\alpha_m, \beta_m > 0$ represents the indices within a start and end period of the area with the useful information;
- $D(k)$ is the current value of the input sample;

- $D'(k)$ means the value of the corresponding output sample focused on the isoelectric line.

Step 7 *The generation of the prototype waveform by averaging a sequence of the cardiac cycles with reduced dissimilarity.* If we assume that we retained K cardiac cycles after the *Step 4* of the processing we can effectuate an average for each sample, with the following relation:

$$F(j) = \frac{1}{K}\sum_{i=1}^{K} D_i'(j), \ (\forall)\, j \in [\alpha_m, \ \beta_m], \tag{6.27}$$

i being the index of the cardiac cycle.
The aim of this averaging is to eliminate a part of the noise, which is considered as a stochastic process, with a null mean value.

Step 8 *The selection of the QRST reduced prototype, which contains the essential information for the IC diagnosis.* One considers that the essential information for the IC diagnosis is contained between the beginning of the Q wave (or the beginning of the R wave if there is not the Q wave) and the end of the T wave, taken on the prototype determined at the *Step 7*. To estimate the initial point of the $QRST$ prototype (the beginning of the Q or R wave) one uses a method similar to that described at the *Step 5*, with $M = 7$.

Step 9 *Normalizing the length of the reduced prototype.* One selects 128 samples that are quasi-uniform spaced on the $QRST$ area. This thing is equivalent to a normalization in the time domain and also constitutes the final step in the statistical processing of the ECG signal.

We have performed the software implementation of our proposed ECG processing cascade [4] and have experimented a new neuro-fuzzy model, entitled Fuzzy Gaussian Neural Network (FGNN) for the IC diagnosis.

We have used [3, 4] an ECG database of 40 subjects, where 20 subjects are IC patients and the other 20 are normal ones. The best performance has been of 100 % IC recognition score. The result is exciting as much as we have used only one lead (V5) of ECG records as input data, while the current diagnosis approaches require the set of 12 lead ECG signals.

We did not use [3, 4] the ECG signals in their natural form, namely as they have been recorded from the respective patients, but in the form of some reduced prototypes, containing 128 samples; Figs. 6.4 and 6.5 show the waveforms of the ECG reduced prototypes, corresponding to the $QRST$ area, both from the ECG of a normal patient and with IC, too.

Figure 6.5 highlights the flattening of the wave T, specific phenomenon for the people with IC.

The ECG signals have to be classified into the following two classes:

(1) class 1 signifies the healthy individuals;
(2) class 2 includes the ill persons.

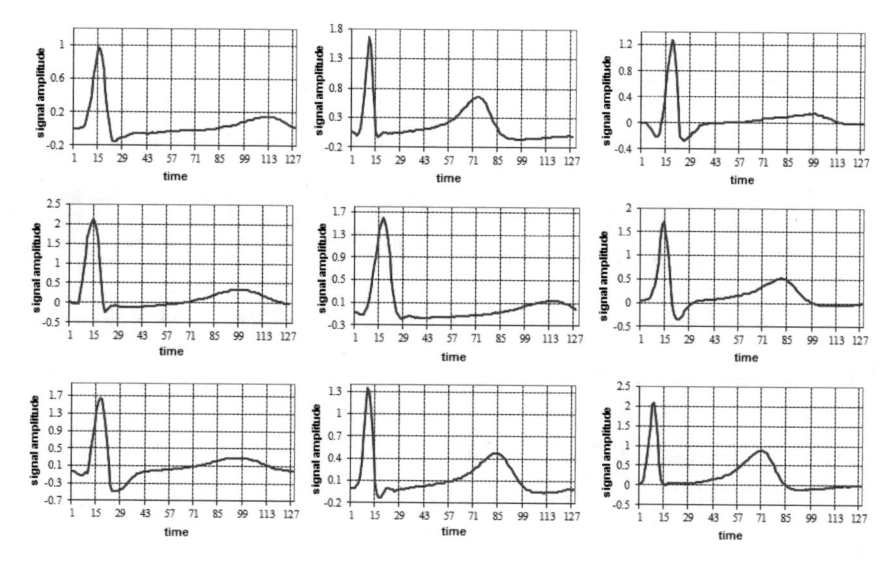

Fig. 6.4 ECG-QRST prototypes corresponding to nine normal patients

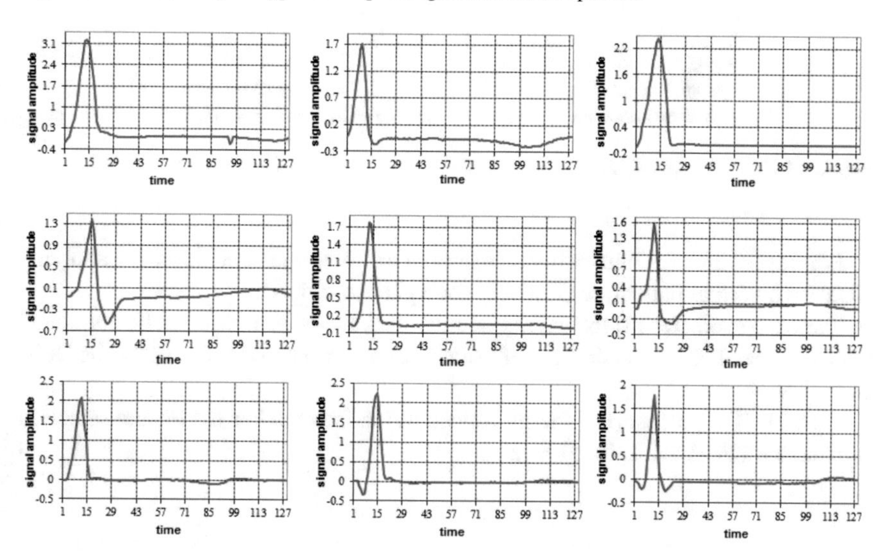

Fig. 6.5 ECG-QRST prototypes corresponding to nine patients afflicted with IC (remark the flat *T* area)

Each of the training and test lot contains 20 reduced prototypes (10 from each class) corresponding to the ECG signals, that belong to the respective subjects.

Figures 6.6 and 6.7 illustrate the vectors in absolute value, achieved by applying PCA for a normal patient, respectively for a patient with IC.

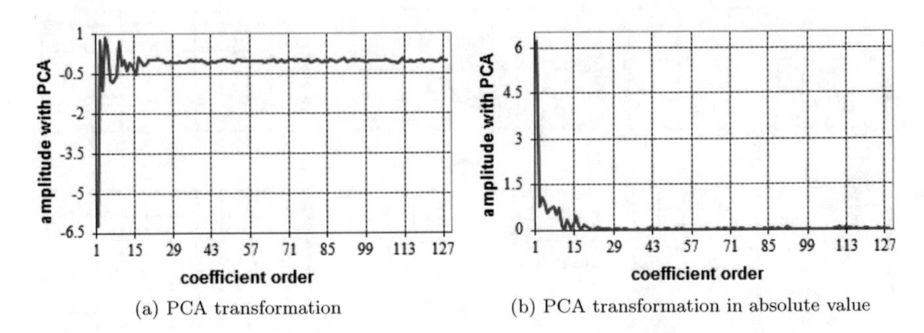

(a) PCA transformation (b) PCA transformation in absolute value

Fig. 6.6 PCA of a normal patient

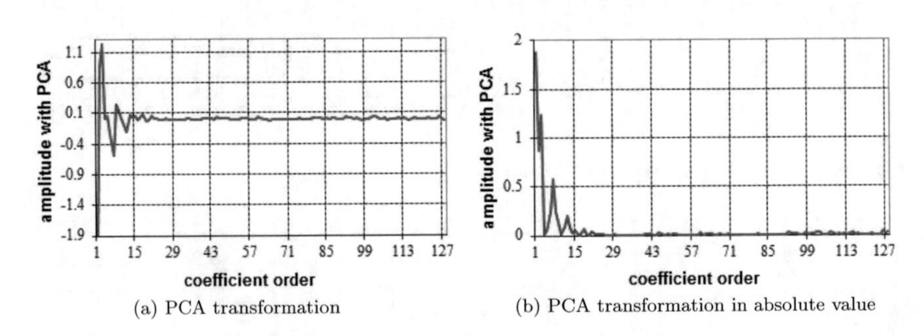

(a) PCA transformation (b) PCA transformation in absolute value

Fig. 6.7 PCA of a patient with IC

Figures 6.8 and 6.9 illustrate the vectors in absolute value, achieved by applying DCT for a normal patient, respectively for a patient with IC.

Table 6.3 evaluates [3] the energy conservation factor F, which can be computed with the formula (6.19) in the case of using PCA and respectively (6.20) for the DCT, assuming $M = 2$.

Figure 6.10 represents the energy concentration factor, when the feature selection is made with PCA/DCT, for $M = 2$.

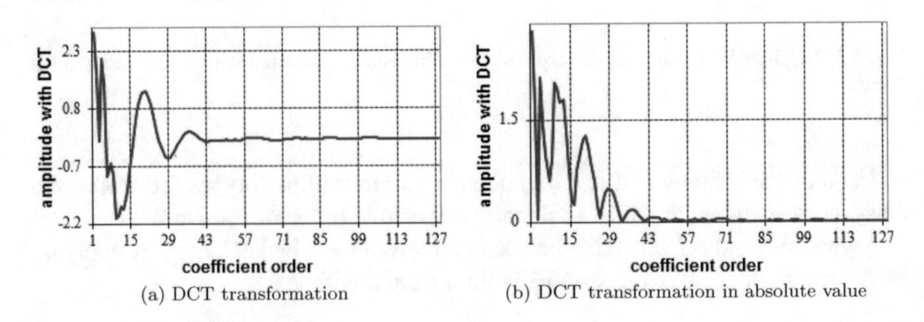

(a) DCT transformation (b) DCT transformation in absolute value

Fig. 6.8 DCT of a normal patient

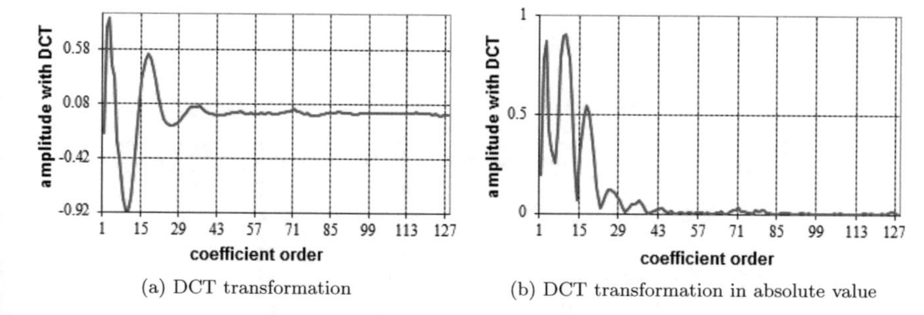

(a) DCT transformation

(b) DCT transformation in absolute value

Fig. 6.9 DCT of a patient with IC

Table 6.3 The energy conservation factor F [%], as a function of m, the number of the features, selected with PCA/DCT ($M = 2$)

m	PCA/DCT	F [%]
28	**PCA**	**100**
	DCT	98.42
30	**PCA**	**100**
	DCT	98.79
38	**PCA**	**100**
	DCT	99.58
40	**PCA**	**100**
	DCT	99.65
48	**PCA**	**100**
	DCT	99.84
50	**PCA**	**100**
	DCT	99.86
58	**PCA**	**100**
	DCT	99.9
60	**PCA**	**100**
	DCT	99.91
68	**PCA**	**100**
	DCT	**99.93**

The DCT requires a less computational complexity than PCA, since it has several fast algorithms available.

By choosing PCA as a feature selection technique, for the training lot of 20 ECG-QRST prototypes (10 normal subjects and 10 afflicted with IC), one can reduce the space dimension from 128 to 28 by preserving 100 % of the signal energy (see Table 6.3).

If one chooses the DCT for the same space dimensionality reduction, the energy preservation ratio decreases to 98.42 % (see Table 6.3).

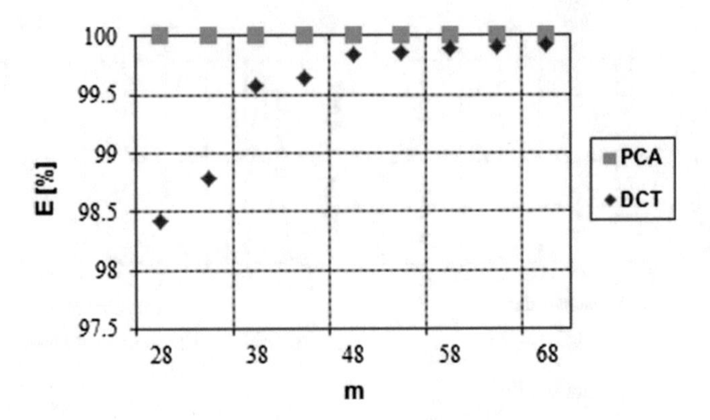

Fig. 6.10 The energy concentration factor ($M = 2$) with PCA/DCT

Table 6.4 Performances of NP versus FNP for ECG classification (M = 2 classes)

Approach	m	n	PCA/DCT	R_1 (%)	R_2 (%)	N_3
NP	28	28	PCA	100	85	3801
			DCT	100	90	2417
	50	50	PCA	100	85	3376
			DCT	100	90	2770
FNP	28	3*28	PCA	100	90	508
			DCT	100	90	218
	50	3*50	PCA	100	90	612
			DCT	**100**	**95**	**239**

Table 6.4 provides [3] the performances of the Nonlinear Perceptron (NP) versus the Fuzzy Nonlinear Perceptron (FNP) for ECG classification (grouped in two classes), with the stage of feature selection (using PCA/DCT), where

- m is the number of features selected using PCA/DCT;
- n signifies the number of input neurons;
- R_1 means the recognition rate (%) corresponding to the training lot (obtained at the end of the training);
- R_2 represents the recognition rate (%) for the test lot (corresponding to the last epoch of training);
- N_3 constitutes the total number of the training epochs.

From Table 6.4 we can note that the best recognition score of 95 %, corresponding to the ECG classification is obtained using the FNP, in the case of making the feature selection with DCT.

Figures 6.11 and 6.12 show [3] the error on the training lot and respectively, the recognition rate over the test lot, obtained with NP/FNP, for $m = 50$, $L = 15$, as a function of the number of epochs.

From Fig. 6.11 we can notice that the error on the training lot decreases much faster when we apply FNP than NP, in the case of the feature selection with PCA/DCT.

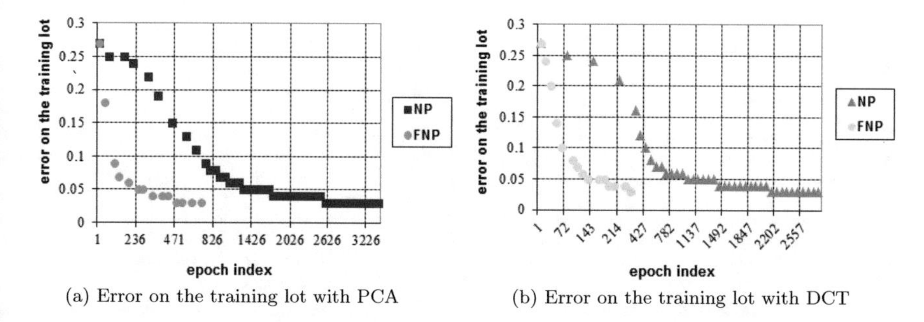

(a) Error on the training lot with PCA

(b) Error on the training lot with DCT

Fig. 6.11 Error on the training lot of NP/FNP with PCA/DCT ($m = 50, L = 15$)

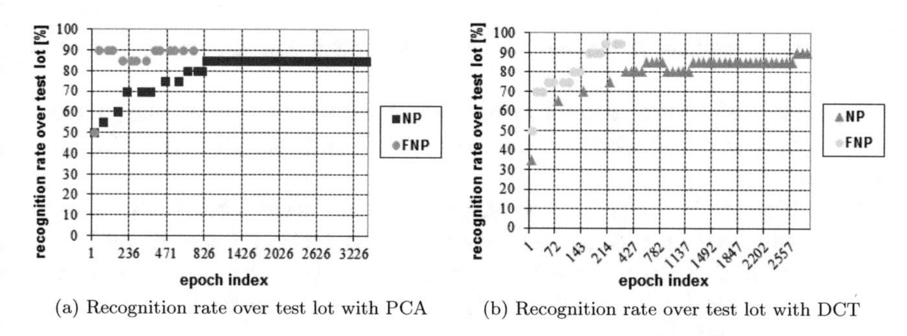

(a) Recognition rate over test lot with PCA

(b) Recognition rate over test lot with DCT

Fig. 6.12 Recognition rate over test lot of NP/FNP with PCA/DCT ($m = 50, L = 15$)

Figure 6.12 proves that the best recognition rates on the test lot is achieved in the case of the feature selection with DCT, both by applying FNP (95 %) and NP (90 %), too.

Figures 6.13 and 6.14 show [3] the error on the training lot and respectively the recognition rate over the test lot, obtained with FNP, in the case of applying PCA/DCT, when $m = 28$ versus $m = 50$, as a function of the number of epochs.

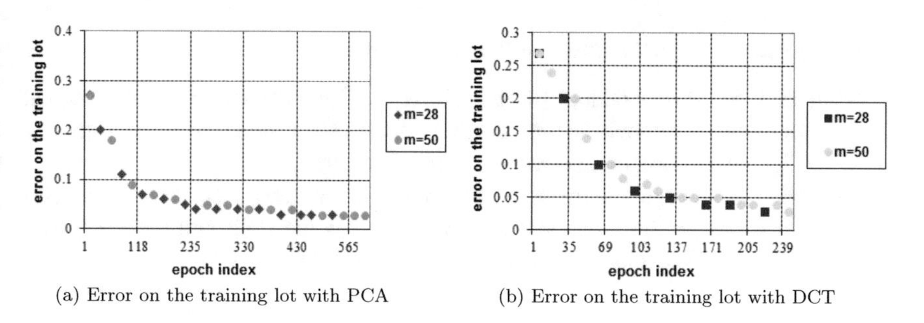

(a) Error on the training lot with PCA

(b) Error on the training lot with DCT

Fig. 6.13 Error on the training lot of FNP with PCA/DCT ($M = 2, L = 15$)

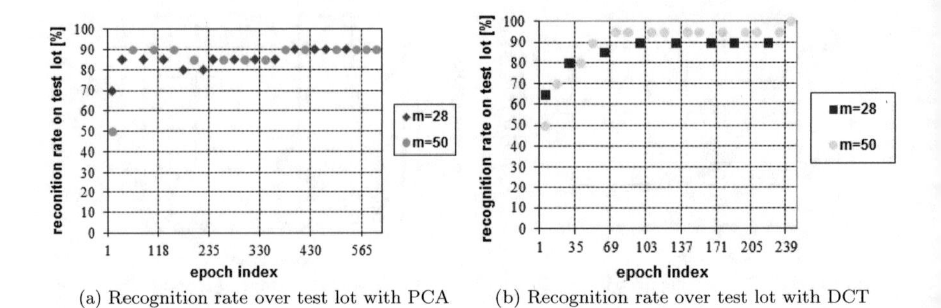

(a) Recognition rate over test lot with PCA (b) Recognition rate over test lot with DCT

Fig. 6.14 Recognition rate over test lot of NP with PCA/DCT ($M = 2, L = 15$)

Usually, by increasing the number of retained features m, the recognition score increases.

Figure 6.13 illustrates a faster decreasing of the error on the training lot corresponding to the FNP when the number of the features selected with PCA/DCT increases.

From Fig. 6.14 it results that FNP determines some better recognition scores on the test lot in case of the feature selection made with the DCT, than with the PCA.

If we use the PCA for the FNP then we shall achieve [3] a recognition rate of 90 % on the test lot, not only when we considered 50 samples but in the case of choosing 28 from the those initial 128.

When we apply DCT, the recognition rate over the test lot obtained using the FNP will be [3] of 95 % in the case of the selection of 50 features and 90 % for 28 selected features.

Table 6.5 provides [3] the performances of the Nonlinear Perceptron (NP) versus the Fuzzy Nonlinear Perceptron (FNP) and (Fuzzy Gaussian Neural Network (FGNN) for the ECG classification (grouped in two classes), with the stage of feature selection (using PCA/DCT, $m = 50$), where:

- R_1 means the recognition rate (%) corresponding to the training lot (obtained at the end of the training);
- R_2 represents the recognition rate (%) for the test lot (corresponding to the last epoch of training);
- N_3 constitutes the total number of the training epochs.

From Table 6.5 we can deduce [3] that the NP, FNP and FGNN needs to be trained in a greater number of epochs in the case of PCA than DCT when we have to classify the ECG signals and obtaining some performant results.

Figure 6.15 delivers the increasing of the recognition rates over the test lot (with the number of the training epochs) of the NP versus the FNP and FGNN, using PCA/DCT for the ECG classification.

Table 6.5 Performances of NP versus FNP and FGNN for ECG classification (M = 2 classes)

Approach	PCA/DCT	R_1 (%)	R_2 (%)	N_3
NP	PCA	100	85	3376
	DCT	100	90	2270
FNP	PCA	100	90	612
	DCT	**100**	**95**	**239**
FNP	**PCA**	**100**	**100**	**1041**
	DCT	100	90	183

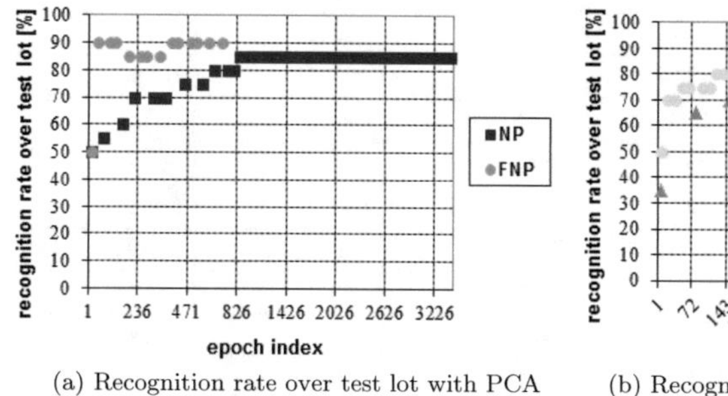

(a) Recognition rate over test lot with PCA

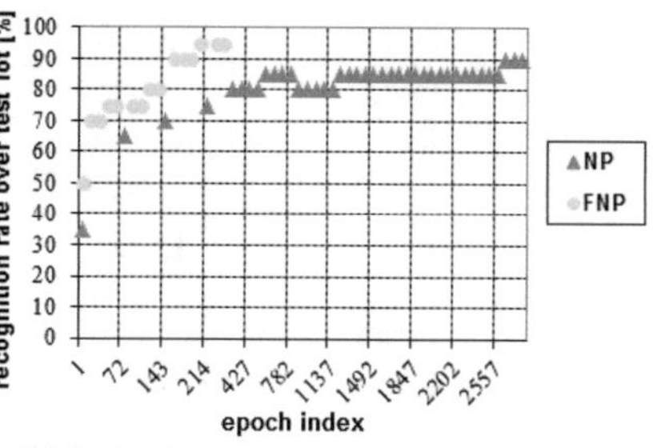

(b) Recognition rate over test lot with DCT

Fig. 6.15 Recognition rate over test lot of NP, FNP, FGNN with PCA/DCT ($M = 2$)

The promising classification performance of the FGNN may be explained [4] by the fact that the classifier is a *hybrid system* of fuzzy logic and a powerful Gaussian network.

In Table 6.5 and Fig. 6.15 one can evaluate [3, 4] the very good recognition performance (100 %) of the FGNN by choosing PCA as a feature extraction stage with $m = 50$ features. The result is exciting as much as we have used only one lead (V5) of ECG records as input data, while the current approaches use the computer processing of 12 lead ECG signals for diagnosis.

From Table 6.5 we conclude that the FGNN training time decreases for DCT by comparison to PCA. For example, choosing $m = 50$ features, the number of necessary training epochs is 1041 for PCA, leading to a recognition rate of 100 % and the number of epochs becomes 183 for DCT, leading to the recognition rate of 90 %.

Table 6.6 shows [3] that for the same number of retained features m, the DCT usually leads to a less recognition rate than PCA (for example, for $m = 50$, one obtains a recognition score of 90 % for DCT and 100 % for PCA; for m = 28, one obtains a recognition score of 90 % for DCT and 95 % for PCA), where:

- m is the number of the features retained using PCA/DCT;
- $\alpha_i,\ i = \overline{1, m},\ 0 < \alpha_i < 1$ are the overlapping factors;

Table 6.6 Performances of FGNN for ECG classification (M = 2 classes), with PCA/DCT

m	PCA/DCT	α_i	R_1 (%)	R_2 (%)	N_3
28	PCA	0.99988	95	90	1713
		0.9999	95	95	1864
		0.99993	95	95	3708
		0.99995	95	95	1868
		0.99997	95	95	2928
	DCT	0.999	100	85	977
		0.9991	100	80	75
		0.9994	**100**	**90**	**2046**
		0.9995	**100**	**90**	**785**
		0.9996	100	80	131
50	PCA	0.99994	100	95	1029
		0.99996	100	95	1048
		0.99997	100	95	647
		0.99998	**100**	**100**	**1041**
		0.99999	100	90	1369
	DCT	**0.99989**	**100**	**90**	**251**
		0.99991	**100**	**90**	**224**
		0.99993	**100**	**90**	**189**
		0.99994	**100**	**90**	**183**
		0.99995	100	90	343

- R_1 constitutes the recognition rate (%) over the training lot (obtained at the end of the training);
- R_2 represents the recognition rate (%) over the test lot (corresponding to the last epoch of training).

The DCT applied for feature extraction has the advantage of reducing the computational effort, but it leads to a slightly less energy-preserving factor by comparison to PCA.

6.4 Concurrent Fuzzy Nonlinear Perceptron Modules

In this section, we shall propose [3, 6] and evaluate a new neuro-fuzzy recognition model called *Concurrent Fuzzy Nonlinear Perceptron Modules* (CFNPM), representing a winner-takes-all collection of small FNP units.

We shall use a following version of a FNP1 (described in the Sect. 1.2.4): if we have vectors of dimension n to be classified, then the input layer of our FNP must have the dimension $3 \cdot n$. Each original component is transformed by fuzzification into three ones, representing the membership of the component for the three linguistic properties: *small*, *medium*, and *big*. The above-mentioned membership functions have been chosen to be expressed by the relations:

$$\mu_{small}(u) = \begin{cases} 1 - \frac{u-0.2}{0.3}, \text{ if } 0.2 \leq u \leq 0.5 \\ 1, \text{ if } 0 \leq u < 0.2 \\ 0, \text{ otherwise} \end{cases} \tag{6.28}$$

$$\mu_{medium}(u) = \begin{cases} 1 - 4 \cdot |u - 0.5|, \text{ if } 0.25 \leq u \leq 0.75 \\ 0, \text{ otherwise} \end{cases} \tag{6.29}$$

$$\mu_{big}(u) = \begin{cases} 1 - \frac{0.8-u}{0.3}, \text{ if } 0.5 \leq u \leq 0.8 \\ 1, \text{ if } 0.8 < u \leq 1 \\ 0, \text{ otherwise} \end{cases} \tag{6.30}$$

that are represented in Fig. 6.16.

The CFNPM system [6] consists of a set of M neuro-fuzzy modules of type FNP, one for each class (see Figs. 6.17 and 6.18).

The significance of the measurements from Fig. 6.18:

- n represents the number of input neurons;
- L means the number of neurons belonging to the hidden layer;
- $X_p = (x_{p1}, \ldots, x_{pn})$ constitutes the input vector;
- W^h_{ji}, $j = \overline{1, L}$, $i = \overline{1, n}$ signifies the set of hidden weights;
- W^o_j, $j = \overline{1, L}$ represents the set of weights corresponding to the output neuron;
- O^k_p, $p = \overline{1, M}$ means the output of the k-th module when the input vector has the index p, where M is the number of neuro-fuzzy modules equal to the number of classes.

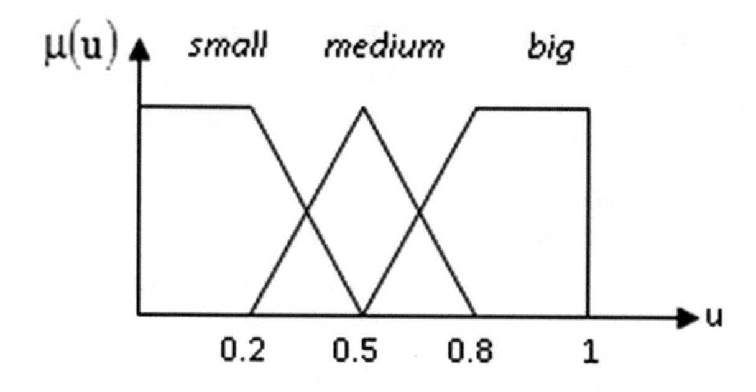

Fig. 6.16 Membership functions for *small*, *medium*, and *big*

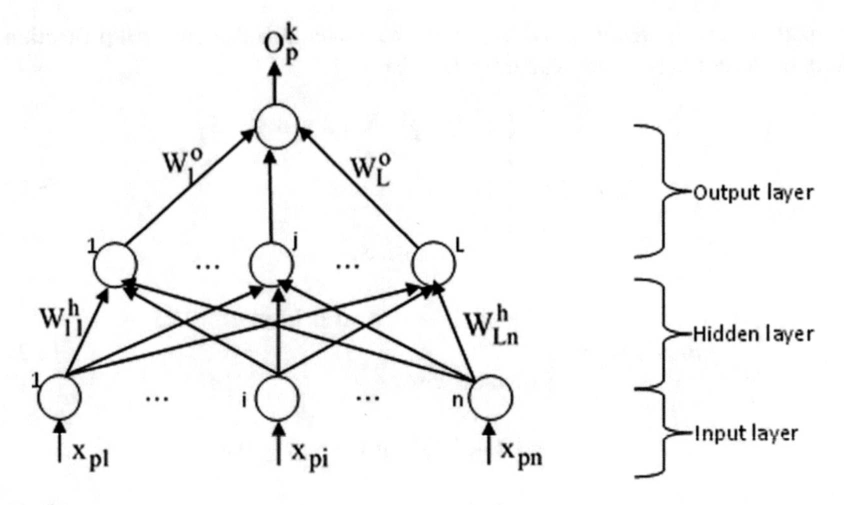

Fig. 6.17 The k-th FNP module

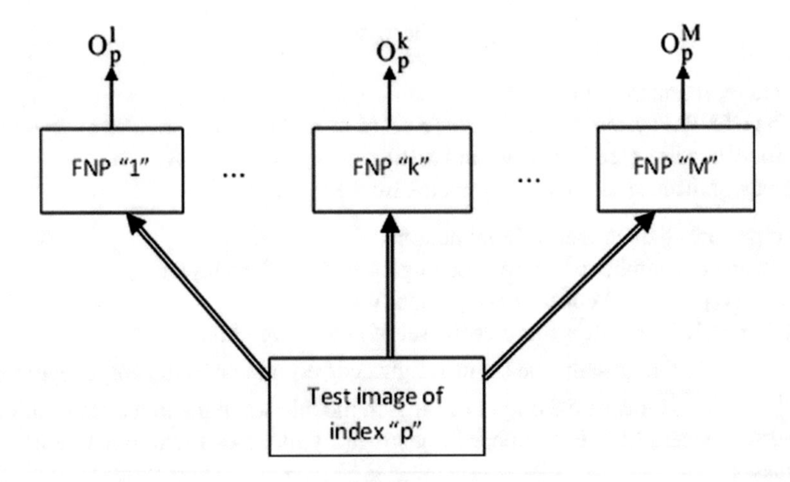

Fig. 6.18 Concurrent Fuzzy Nonlinear Perceptron Modules (CFNPM)

The ideal output of the k-th ($k = \overline{1, M}$) FNP module is equal to 1, for the training images belonging to the class k (the number of the images from each class being $N_1 = \cdots = N_k = \cdots = N_M = N_c$) and is equal to 0, for the other images of the training lot.

The training lot corresponding to the network of the index k consists in $(M-1) \cdot N_c$ images (and it differs from one network to another), where:

- a half of them represents a half of the images with the label k, repeated $(M - 1)$ times,
- the rest of $(M - 1) \cdot \frac{N_c}{2}$ images result from the application of $\frac{N_c}{2}$ images for each of the other $(M - 1)$ classes.

After the M neural networks were trained in a number of epochs, the weights were "frozen" and saved in M files to be used in the test stage of the respective networks.

The test lot consists in $N_t = M \cdot \frac{N_c}{2}$ images, that are different from those used in the learning algorithm (by $\frac{N_c}{2}$ images from every class).

The classification phase consists of the following steps:

(a) one applies the test image to the input of each of the M modules and computes the corresponding outputs;
(b) evaluate the maximum of the M outputs computed at step (a);
(c) associate to the input image the class label of the neuro-fuzzy module that leads to the maximum output.

6.5 Concurrent Fuzzy Gaussian Neural Network Modules

Our fuzzy neural model [3, 7], entitled *Concurrent Fuzzy Gaussian Neural Network Modules* (CFGNNM) (see Fig. 6.19) consists of a set of M fuzzy neural networks, by the type FGNN (Fuzzy Gaussian Neural Network), one for every class, each network from Fig. 6.20 having a single output.

The meaning of the measurements from Fig. 6.19:

- m represents the number of the neurons corresponding to the input layer;
- K signifies the number of the neurons from the third layer;
- M is the number of the neuro-fuzzy modules equal to the number of classes;
- $X = (x_1, \ldots, x_m)$ constitutes the input vector;

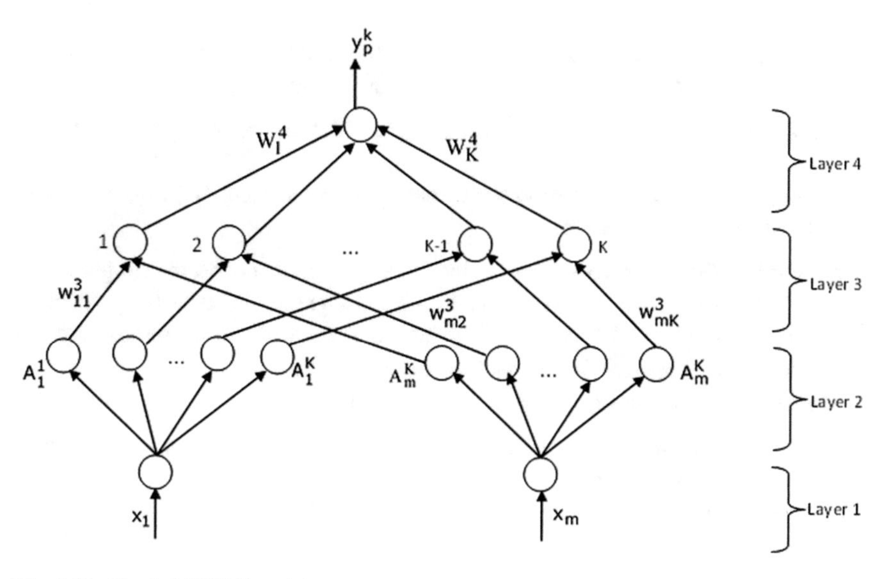

Fig. 6.19 The k-th FGNN module

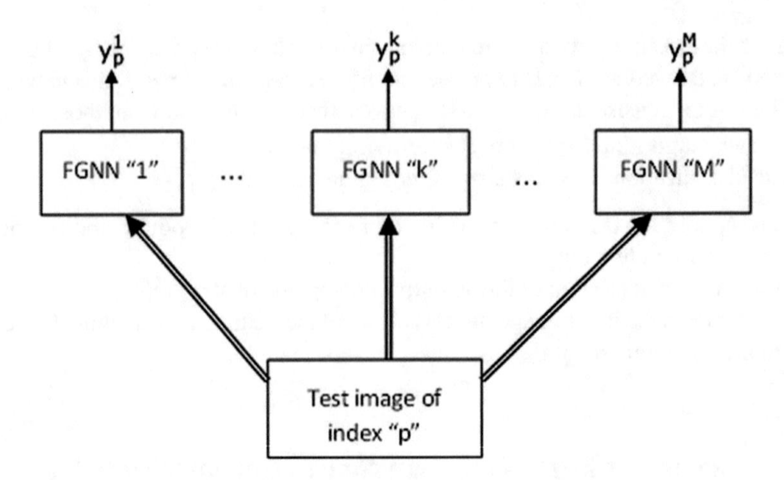

Fig. 6.20 Concurrent Fuzzy Gaussian Neural Network Modules (CFGNNM)

- W_{ij}^3, $i = \overline{1, m}$, $j = \overline{1, K}$ is the weight between the $(i - 1)k + j$-th neuron of the second layer and the neuron j of the third layer;
- W_i^4, $i = \overline{1, K}$ represents the connection from the neuron i from the third layer and the neuron from the last layer of the FGNN module;
- y_p^k, $k = \overline{1, M}$ means the output of the k-th module, when the input vector has the index p.

The ideal output of the k-th ($k = \overline{1, M}$) FGNN module is equal to 1, for the training images belonging to the class k (the number of the images from each class being $N_1 = \cdots = N_k = \cdots = N_M = N_c$) and is equal to 0, for the rest of the training lot.

The training lot corresponding to the network of the index k consists in $(M-1) \cdot N_c$ images (and it differs from one network to another), where:

- a half of them represents a half of the images with the label k, repeated $(M - 1)$ times,
- the rest of $(M - 1) \cdot \frac{N_c}{2}$ images result from the application of $\frac{N_c}{2}$ images for each of the other $(M - 1)$ classes.

After the M neural networks were trained in a number of epochs, the weights were "frozen" and saved in M files to be used in the test stage of the respective networks.

The test lot consists of $N_t = M \cdot \frac{N_c}{2}$ images, that are different from those used in the learning algorithm (by $\frac{N_c}{2}$ images from every class).

The classification phase consists of the following steps:

(a) one applies the test image to the input of each of the M modules and computes the corresponding outputs;
(b) evaluate the maximum of the M outputs computed at step (a);
(c) associate to the input image the class label of the neuro-fuzzy module that leads to the maximum output.

6.6 Experimental Results

We shall experiment the following five approaches:

(1) Nonlinear Perceptron (NP),
(2) Fuzzy Nonlinear Perceptron (FNP),
(3) Fuzzy Gaussian Neural Network (FGNN),
(4) Concurrent Fuzzy Nonlinear Perceptron Modules (CFNPM),
(5) Concurrent Fuzzy Gaussian Neural Network Modules (CFGNNM),

using "The ORL Database of Faces" provided by the AT&T Laboratories from Cambridge University, with 400 images, corresponding to 40 subjects (namely, 10 images for each class).

We have implemented the software corresponding to the presented neural models for face recognition application using Microsoft Visual C++.

For feature extraction, we have chosen the Principal Component Analysis (PCA) technique versus Discrete Cosine Transform (DCT).

To achieve the selection stage of feature selection with PCA and DCT respectively, the original images (portraits) of size 92×112 have been reduced to size 46×56. We considered the situation of $M = 40$ classes of subjects and 400 images (each of 40 subjects appear in 10 hypostases). From these images have resulted the vectors $X_i,\ i = \overline{1, 400}$, having the size $45 \times 56 = 2576$.

The experimental results are given in Table 6.7, which provides the performances of the NP, FNP, FGNN versus CFNPM and CFGNNM systems, to face recognition (grouped in 40 classes), with the stage of feature selection (using PCA/DCT), where:

- m is the number of selected features using PCA/DCT;
- n signifies the number of input neurons;
- R_1 means the recognition rate (%) corresponding to the training lot (obtained at the end of the training);
- R_2 represents the recognition rate (%) for the test lot (corresponding to the last epoch of training).

In the case of using NP, FNP and CFNPM in our experiment, we considered that the number L of neurons belonging to the hidden layer is equal to 200.

The results emphasized in Table 6.7 prove that the use of the two concurrent neural networks: CFNPM by the type FNP and respectively CFGNNM by the type FGNN causes an increase in the recognition rates for the training and the test lot (both in the case of making the feature selection with PCA and DCT, too) compared to those achieved using the simple variations of FNP and FGNN.

From Table 6.7, we can evaluate that:

- for $m = 100$, the best recognition score over the test lot (91.50 %) is obtained by the CFGNNM (with DCT);
- the best recognition performance (93 %) is obtained using a cascade of PCA (m = 160)—CFNPM.

Table 6.7 Performances of NP, FNP, FGNN versus CFNPM and CFGNNM for face recognition (M = 40 classes)

Approach	PCA/DCT	m	n	R_1 (%)	R_2 (%)
NP	PCA	100	100	99	88
		160	160	99	88.50
	DCT	100	100	99	89
		160	160	99	89
FNP	PCA	100	3 x 100	99.50	85
		160	3 x 160	100	91
	DCT	100	3 x 100	97.50	84.50
		160	3 x 160	99.50	88.50
FGNN	PCA	100	100	99	77
		160	160	100	84
	DCT	100	100	100	91
		160	160	100	91
CFNPM	PCA	100	3 x 100	100	86.50
		160	**3 x 160**	**100**	**93**
	DCT	100	3 x 100	99.50	86
		160	3 x 160	100	89
CFGNNM	PCA	100	100	100	79
		160	160	100	84.50
	DCT	**100**	**100**	**100**	**91.50**
		160	160	100	91.50

Table 6.8 Comparison between the performances of the FGNN and CFGNNM systems experimented to the face recognition ($M = 40$) with the feature selection

Approach	m	PCA/DCT	α_i	R_1 (%)	R_2 (%)
FGNN	100	PCA	0.999999	99	77
CFGNNM	100			100	79
FGNN	160		0.999999	100	84
CFGNNM	160			**100**	**84.50**
FGNN	100	DCT	0.9999991	100	91
CFGNNM	100			**100**	**91.50**
FGNN	160		0.9999994	100	91
CFGNNM	160			**100**	**91.50**

Table 6.8 shows that the FGNN and CFGNNM approaches are sensitive to the changing of the values corresponding to the overlapping factors, defined in equation (3.18), where:

- m is the number of selected features using PCA/DCT;
- $\alpha_i,\ i = \overline{1, m},\ 0 < \alpha_i < 1$ are the overlapping factors;
- R_1 means the recognition rate (%) for the training lot (obtained at the end of the training);
- R_2 represents the recognition rate (%) for the test lot (corresponding to the last epoch of training).

From Table 6.8 is evaluated that the best recognition performance (91.50 %) is achieved using a cascade of DCT (both for $m = 100$ and $m = 160$)—CFGNNM.

We can also noticed that the use of the Concurrent Fuzzy Gaussian Neural Network causes an increase of the recognition rates (even when we change the values of the overlapping factors) for the training lot and the test lot (in case of using PCA/DCT) compared with the those obtained using the simple variant of FGNN.

References

1. V. E. Neagoe and O. Stănăşilă. *Theories of Pattern Recognition (in Romanian)*. Publishing House of the Romanian Academy, Bucharest, 1992.
2. R.C. Gonzales and A. Woods. *Digital Image Processing*. Prentice Hall, second edition, 2002.
3. I. Iatan. *Neuro- Fuzzy Systems for Pattern Recognition (in Romanian)*. PhD thesis, Faculty of Electronics, Telecommunications and Information Technology- University Politehnica of Bucharest, PhD supervisor: Prof. dr. Victor Neagoe, 2003.
4. V. Neagoe, I. Iatan, and S. Grunwald. A neuro- fuzzy approach to ecg signal classification for ischemic heart disease diagnosis. In *the American Medical Informatics Association Symposium (AMIA 2003), Nov. 8- 12 2003, Washington DC*, pages 494–498, 2003.
5. V. E. Neagoe and O. Stănăşilă. *Pattern Recognition and Neural Networks (in Romanian)*. Ed. Matrix Rom, Bucharest, 1999.
6. V. Neagoe and I. Iatan. Concurrent fuzzy nonlinear perceptron modules for face recognition. In *Proceedings of the International Conference COMMUNICATIONS 2004*, pages 269–274. Edited by Technical Military Academy, Bucharest, 2004.
7. I. Iatan. A concurrent fuzzy neural network approach for a fuzzy gaussian neural network. *Blucher Mechanical Engineering Proceedings*, 1(1):3018–3025, 2014.

Chapter 7
A New Interval Arithmetic-Based Neural Network

The aim of this chapter is to design a new model of fuzzy nonlinear perceptron, based on alpha level sets. The new model entitled *Fuzzy Nonlinear Perceptron based on Alpha Level Sets* (FNPALS) [1, 2] differs from the other fuzzy variants of the nonlinear perceptron, where the fuzzy numbers are represented by membership values. In the case of FNPALS, the fuzzy numbers are represented throught the alpha level sets.

7.1 The Representation of the Fuzzy Numbers

From [3, 4] it is known the use of the intervals for representing uncertain inputs and missing data.

An interval can be represented by its lower limit and upper limit as

$$A = [a^L, a^R],$$

where the superscripts L and R denote the lower limit (left) and the upper limit (right), respectively.

The following addition and multiplication are used in this chapter for calculating the total input to each unit in interval- arithmetic-based neural networks:

$$A + B = [a^L, a^R] + [b^L, b^R] = [a^L + b^L, a^R + b^R],$$

$$a \cdot B = a \cdot [b^L, b^R] = \begin{cases} \left[a \cdot b^L, a \cdot b^R\right], & \text{if } a \geq 0 \\ \left[a \cdot b^R, a \cdot b^L\right], & \text{if } a < 0 \end{cases}$$

$$A \cdot B = [a^L, a^R] \cdot [b^L, b^R] = [\min\{a^L b^L, a^L b^R, a^R b^L, a^R b^R\}, \max\{a^L b^L, a^L b^R, a^R b^L, a^R b^R\}].$$

I.F. Iatan, *Issues in the Use of Neural Networks in Information Retrieval*, Studies in Computational Intelligence 661, DOI 10.1007/978-3-319-43871-9_7

In the case of $0 \leq a^L \leq a^R$ (i.e., if A is nonnegative) the preceding product operation on intervals can be simplified as

$$A \cdot B = [a^L, a^R] \cdot [b^L, b^R] = [\min\{a^L b^L, a^R b^L\}, \max\{a^L b^R, a^R b^R\}].$$

"Interval representation is also useful for utilizing experts knowledge in the learning of neural networks".[1]

We want [2] to improve the fuzzy neural model from [3], where has proposed a one-layer perceptron for classification tasks, where inputs and outputs are represented by intervals. As in the case of the MLP (Multilayer Perceptron), an interval Multilayer Perceptron (iMLP) with n inputs and m outputs is comprised of:

- an input layer with n input buffer units;
- one or more hidden layers with a nonfixed number of nonlinear hidden units;
- one output layer with m linear or nonlinear output units.

The model from [3] has restricted to one hidden layer with h hidden neurons and one output ($m = 1$). The iMLP described below is comprised [3] of two layers of adaptive weights (a hidden layer and an output layer).

Considering n interval-valued inputs

$$X_i = < x_i^C, x_i^R > = [x_i^C - x_i^R, x_i^C + x_i^R], \ i = \overline{1, n},$$

the output of the j-th hidden unit is achieved by forming a weighted linear combination of the n interval inputs and the bias. As the weights of the proposed architecture are crisp and not intervals, this linear combination results in a new interval given by:

$$S_j = w_{j0} + \sum_{i=1}^{n} w_{ji} X_i = \left\langle w_{j0} + \sum_{i=1}^{n} w_{ji} x_i^C, \sum_{i=1}^{n} |w_{ji}| x_i^R \right\rangle. \tag{7.1}$$

The activation of the hidden unit j will be obtained by transforming the interval S_j using a nonlinear activation function $g()$:

$$A_j = g(S_j). \tag{7.2}$$

In [3], the tanh function is used as an activation function in the hidden layer. As the activation function is monotonic, this transformation yields to a new interval; it can be calculated as:

$$A_j = \tanh(S_j) = [\tanh(s_j^C - s_j^R), \tanh(s_j^C + s_j^R)] =$$

$$= \left\langle \frac{\tanh(s_j^C - s_j^R) + \tanh(s_j^C + s_j^R)}{2}, \frac{\tanh(s_j^C + s_j^R) - \tanh(s_j^C - s_j^R)}{2} \right\rangle. \tag{7.3}$$

[1] Leondes, C. T., Fuzzy Logic and Expert Systems Applications. San Diego, Academic Press, 1998.

Finally, the output of the network, $\hat{Y}$ is achieved by transforming the activations of the hidden units by a second layer of processing units. In the case of a single output and a linear activation function with crisp weights, the estimated output interval is achieved as a linear combination of the activations of the hidden layer and the bias:

$$\hat{Y} = \sum_{j=1}^{h} \alpha_j A_j + \alpha_0 = \left\langle \sum_{j=1}^{h} \alpha_j a_j^C + \alpha_0, \sum_{j=1}^{h} |\alpha_j| a_j^R \right\rangle, \tag{7.4}$$

but its transfer function has been updated in order to operate with interval-valued inputs and outputs.

The old model has been used as an interval-valued function approximation model, whose crisp weights can be adjusted using a supervised learning procedure by minimizing an error function of the form:

$$E = \frac{1}{P} \sum_{t=1}^{P} d(Y(t), \hat{Y}(t)) + \lambda \Phi(\hat{f}), \tag{7.5}$$

where:

- $d(Y(t), \hat{Y}(t))$ is a measure of the discrepancy between the desired and the estimated output intervals for the t-th training sample, denoted by $Y(t)$ and $\hat{Y}(t)$, respectively;
- $\lambda \Phi(\hat{f})$ is a regularization term of the estimated function $\hat{f}(X) : X \to Y$.

The resulting model maps an input of intervals to an interval output by means of interval arithmetic.

The fuzzy number can be represented using the following two methods:

(1) the method proposed by Umano and Ezawa [5], according to which, a fuzzy number is represented by a finite number of membership values;
(2) the method of Uehara and Fujise [6], regarding the representation of a fuzzy number by a finite number of alpha level sets.

Let be a set of K fuzzy rules, each rule being of the form:

$$R_p : \text{If } x_1 \text{is } A_1^p \text{ and } x_2 \text{is } A_2^p \text{ and} \ldots \text{and } x_n \text{is } A_n^p \text{ then } y_1 \text{is } B_1^p \text{ and } y_2 \text{is } B_2^p \text{ and} \ldots \text{and } y_m \text{is } B_m^p, \tag{7.6}$$

$p = \overline{1, K}$, where:

- $A_i^p \; i = \overline{1, n}$ and $B_j^p \; j = \overline{1, m}$ are some fuzzy numbers;
- n is number of the terms from the antecedent side of the rule p;
- m means the number of the terms which appears in the consequent of the rule p.

The set of the input/ output pairs of the fuzzy system are:

$$\{(A_1^p, A_2^p, \ldots, A_n^p), (B_1^p, B_2^p, \ldots, B_m^p)\}. \tag{7.7}$$

7.1.1 Representing the Fuzzy Numbers by a Finite Number of Membership Values

Let $[\alpha_1, \alpha_2]$ be an interval which contains the support corresponding to the inputs of a fuzzy system and $[\beta_1, \beta_2]$, the interval which contains the support corresponding to the outputs.

One assumes:

$$\begin{cases} x_j = \alpha_1 + (j-1) \cdot \frac{\alpha_2-\alpha_1}{n-1}, \; j = \overline{1,n} \\ y_i = \beta_1 + (i-1) \cdot \frac{\beta_2-\beta_1}{m-1}, \; i = \overline{1,m}, \end{cases} \tag{7.8}$$

where $m \geq 2, \; n \geq 2$.

A discrete version of the training set (7.7) is the set of the input/output pairs:

$$\left\{ \left(A_1^p(x_1), \; A_2^p(x_2), \ldots, A_n^p(x_n)\right), \; \left(B_1^p(y_1), \; B_2^p(y_2), \ldots, B_m^p(y_m)\right) \right\}. \tag{7.9}$$

Example 7.1 We shall consider the following fuzzy rule-based system, consisting in the three rules:

R_1 : If x is *small* then y is *negative*.

R_2 : If x is *medium* then y is *about zero*.

R_3 : If x is *big* then y is *positive*.

The membership functions of the fuzzy terms:

- *small*, *medium*, and *big* were defined in the relations (6.28), (6.29), and respectively (6.30) and represented in Fig. 6.16;
- *negative*, *about zero*, and *positive* are:

$$\mu_{negative}(u) = \begin{cases} -u, \text{ if } -1 \leq u \leq 0 \\ 0, \text{ otherwise} \end{cases} \tag{7.10}$$

$$\mu_{about\ zero}(u) = \begin{cases} 1 - 2 \cdot |u|, \text{ if } -\frac{1}{2} \leq u \leq \frac{1}{2} \\ 0, \text{ otherwise} \end{cases} \tag{7.11}$$

$$\mu_{big}(u) = \begin{cases} u, \text{ if } 0 \leq u \leq 1 \\ 0, \text{ otherwise} \end{cases} \tag{7.12}$$

being illustrated in Fig. 7.1.

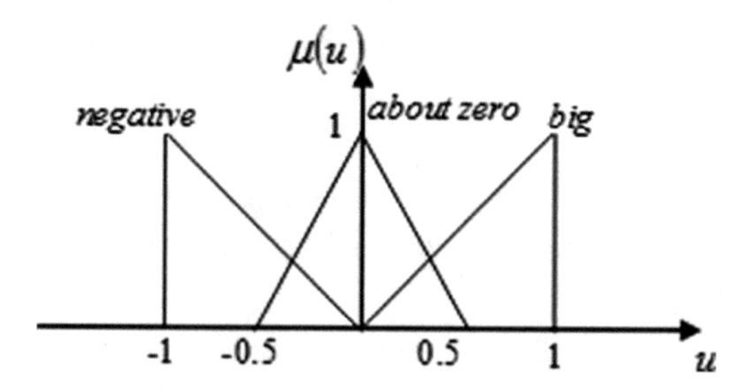

Fig. 7.1 Membership functions for *negative*, *about zero*, and *positive*

The training set derived from the fuzzy rule-based system described the Example 7.1 can be written in the form:

$$\{(small,\ negative),\ (medium,\ about\ zero),\ (big,\ positive)\}. \tag{7.13}$$

Let [0, 1] be the interval which contains the support corresponding to the inputs of the considered fuzzy system and $[-1,\ 1]$, the interval which contains the support corresponding to the outputs.

We assume $m = n = 5$ and:

$$\begin{cases} x_j = \frac{j-1}{4},\ j = \overline{1,n} \\ y_i = -1 + (i-1) \cdot \frac{2}{4} = -1 + i - \frac{1}{2} = -\frac{3}{2} + \frac{i}{2},\ i = \overline{1,m}. \end{cases}$$

We shall compute:

$$x_1 = 0,\quad x_2 = 0.25,\ x_3 = 0.5,\ x_4 = 0.75,\ x_5 = 1$$
$$y_1 = -1,\ y_2 = -0.5,\ y_3 = 0.5,\ y_4 = 0.5,\ y_5 = 1.$$

The discrete version of the training continuous lot consists in the three input/output pairs:

$$\{(a_{11},\ a_{12}, \ldots, a_{15}),\ (b_{11},\ b_{12}, \ldots, b_{15})\},$$
$$\{(a_{21},\ a_{22}, \ldots, a_{25}),\ (b_{21},\ b_{22}, \ldots, b_{25})\},$$
$$\{(a_{31},\ a_{32}, \ldots, a_{35}),\ (b_{31},\ b_{32}, \ldots, b_{35})\},$$

where

$$\begin{cases} a_{1j} = \mu_{small}(x_j),\ (\forall)\, j = \overline{1,5} \\ a_{2j} = \mu_{medium}(x_j),\ (\forall)\, j = \overline{1,5} \\ a_{3j} = \mu_{big}(x_j),\ (\forall)\, j = \overline{1,5} \\ b_{1i} = \mu_{negative}(y_i),\ (\forall)\, i = \overline{1,5} \\ b_{2i} = \mu_{about\ zero}(y_i),\ (\forall)\, i = \overline{1,5} \\ b_{3i} = \mu_{positive}(y_i),\ (\forall)\, i = \overline{1,5}. \end{cases}$$

By passing to the numerical values we achieve the training set from (7.13).

7.1.2 *Representing the Fuzzy Numbers by a Finite Number of Alpha level Sets*

In the case of considering this method for the representation of the fuzzy number one chooses a natural number $M \geq 2$, which represents the number of the α level sets and one supposes

$$\alpha_j = \frac{j-1}{M-1},\ j = \overline{1, M}, \tag{7.14}$$

a partition of the interval [0, 1].

We shall apply the present method for the fuzzy rule database described in (7.6).

The M alpha level sets corresponding to the fuzzy number $A_i^p,\ i = \overline{1, n},\ p = \overline{1, K}$ are

$$\left[A_i^p\right]^{\alpha_j} = \left\{u | A_i^p(u) \geq \alpha_j\right\} = \left[a_{pij}^L,\ a_{pij}^R\right],\ j = \overline{1, M},\ i = \overline{1, n},\ p = \overline{1, K}.$$

Similarly, by definition, the M alpha level sets of the fuzzy number $B_i^p,\ i = \overline{1, m},\ p = \overline{1, K}$ are given by the relation:

$$\left[B_i^p\right]^{\alpha_j} = \left\{u | B_i^p(u) \geq \alpha_j\right\} = \left[b_{pij}^L,\ b_{pij}^R\right],\ j = \overline{1, M},\ i = \overline{1, m},\ p = \overline{1, K}.$$

The set of the input/output pairs corresponding to the fuzzy systems can be written using the number M of the α level sets in the form:

$$\left\{ \left(a_{pi1}^L, a_{pi1}^R \ldots, a_{piM}^L, a_{piM}^R\right),\ \left(b_{pj1}^L, b_{pj1}^R \ldots, b_{pjM}^L, b_{pjM}^R\right) \right\}, \tag{7.15}$$

$i = \overline{1, n},\ j = \overline{1, m},\ p = \overline{1, K}.$

Theorem 7.1 ([4]). *For a fuzzy triangular number, represented in Fig. 7.2, having the general form*

$$A(t) = \begin{cases} 1 - \frac{a-t}{\gamma}, \text{ if } a - \gamma \leq t \leq a \\ 1 - \frac{t-a}{\tau}, \text{ if } a < t \leq a + \tau \\ 0, \text{ otherwise} \end{cases} \tag{7.16}$$

the α level sets are computed with the formula:

$$[A]^\alpha = [a - (1-\alpha)\gamma,\ a + (1-\alpha)\tau]. \tag{7.17}$$

Figure 7.3 illustrates the representation of a fuzzy triangular number by M alpha level sets.

Theorem 7.2 ([4]) *In the case of a fuzzy trapezoidal number, represented in Fig. 7.4, which has the general form*

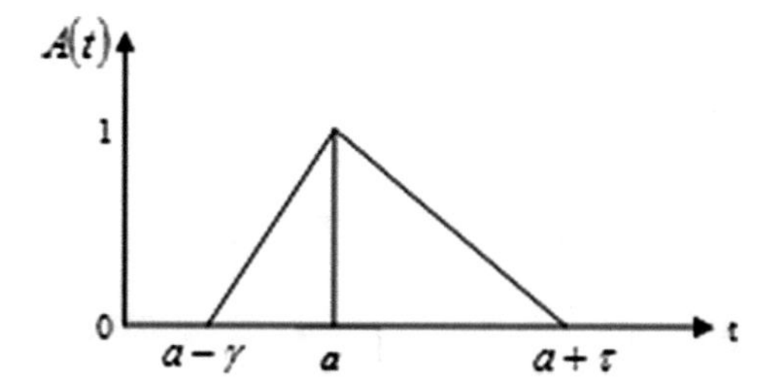

Fig. 7.2 A fuzzy triangular number

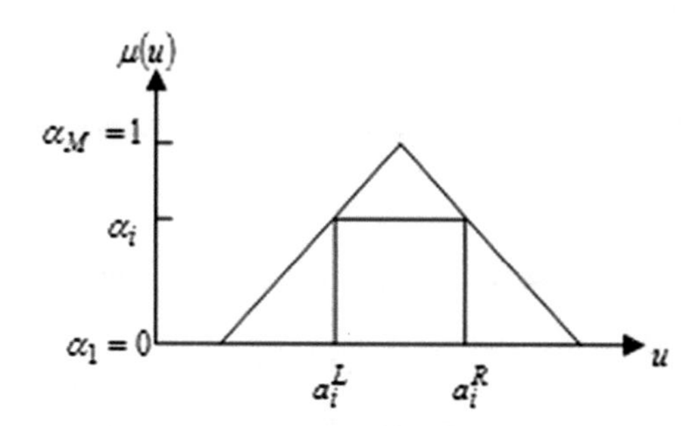

Fig. 7.3 The representation of a fuzzy triangular number by M alpha level sets

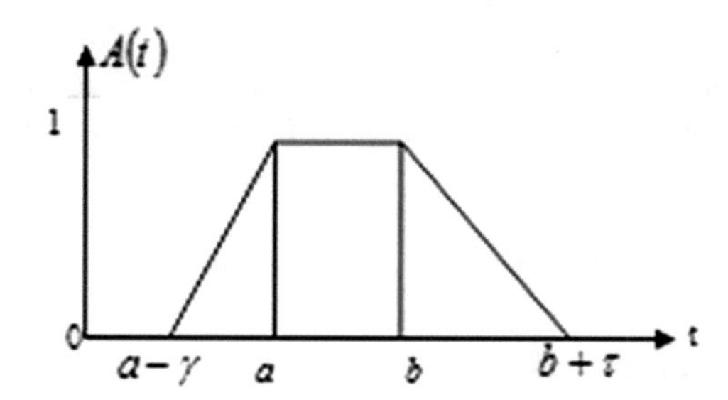

Fig. 7.4 A fuzzy trapezoidal number

$$A(t) = \begin{cases} 1 - \frac{a-t}{\gamma}, \text{ if } a - \gamma \leq t < a \\ 1, \text{ if } a \leq t \leq b \\ 1 - \frac{t-b}{\tau}, \text{ if } b < t \leq b + \tau \\ 0, \text{ otherwise} \end{cases} \tag{7.18}$$

we can compute the α *level sets using the formula:*

$$[A]^{\alpha} = [a - (1-\alpha)\gamma,\ b + (1-\alpha)\tau]. \tag{7.19}$$

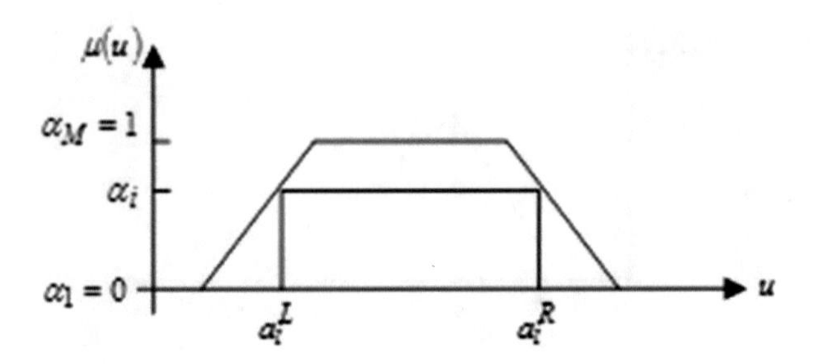

Fig. 7.5 The representation of a fuzzy trapezoidal number by M alpha level sets

Figure 7.5 shows the representation way of a fuzzy trapezoidal number by M alpha level sets.

Example 7.2 We assume a fuzzy rule-based system, which consists in the the following three rules:

$$R_1 : \text{ If } x \text{ is } small \text{ then } y \text{ is } small.$$

$$R_2 : \text{ If } x \text{ is } medium \text{ then } y \text{ is } medium.$$

$$R_3 : \text{ If } x \text{ is } big \text{ then } y \text{ is } big.$$

The membership functions of the fuzzy terms *small*, *medium*, and *big* are defined in the relations (6.28), (6.29), and respectively (6.30) and represented in Fig. 6.16.

Let

$$\alpha_j = \frac{j-1}{M-1}, \; j = \overline{1, M}, \; M = 6 \tag{7.20}$$

be a partition of the interval [0, 1].

We shall achieve:

$$\alpha_1 = 0, \alpha_2 = 0.2, \; \alpha_3 = 0.4, \; \alpha_4 = 0.6, \; \alpha_5 = 0.8, \; \alpha_6 = 1.$$

The set of the input/output pairs corresponding to the fuzzy systems will be:

$$\begin{cases} \left\{ \left(a_{11}^L, a_{11}^R \ldots, a_{16}^L, a_{16}^R\right), \; \left(b_{11}^L, b_{11}^R \ldots, b_{16}^L, b_{16}^R\right) \right\} \\ \left\{ \left(a_{21}^L, a_{21}^R \ldots, a_{26}^L, a_{26}^R\right), \; \left(b_{21}^L, b_{21}^R \ldots, b_{26}^L, b_{26}^R\right) \right\} \\ \left\{ \left(a_{31}^L, a_{31}^R \ldots, a_{36}^L, a_{36}^R\right), \; \left(b_{31}^L, b_{31}^R \ldots, b_{36}^L, b_{36}^R\right) \right\}, \end{cases} \tag{7.21}$$

where:

$$\begin{cases} \left[a_{1j}^L,\ a_{1j}^R\right] = \left[b_{1j}^L,\ b_{1j}^R\right] = [\text{small}]^{\alpha_j} \\ \left[a_{2j}^L,\ a_{2j}^R\right] = \left[b_{2j}^L,\ b_{2j}^R\right] = [\text{medium}]^{\alpha_j} \\ \left[a_{3j}^L,\ a_{3j}^R\right] = \left[b_{3j}^L,\ b_{3j}^R\right] = [\text{big}]^{\alpha_j}. \end{cases} \tag{7.22}$$

Substituting with numerical values in the relation (7.21), we achieve:

$$\begin{cases} \{(0,\ 0.5,\ 0,\ 0.44,\ 0,\ 0.38,\ 0,\ 0.32,\ 0,\ 0.26,\ 0,\ 0.2), \\ (0,\ 0.5,\ 0,\ 0.44,\ 0,\ 0.38,\ 0,\ 0.32,\ 0,\ 0.26,\ 0,\ 0.2)\} \\ \{(0.25,\ 0.75,\ 0.3,\ 0.7,\ 0.35,\ 0.65,\ 0.4,\ 0.6,\ 0.45,\ 0.55,\ 0.5,\ 0.5), \\ (0.25,\ 0.75,\ 0.3,\ 0.7,\ 0.35,\ 0.65,\ 0.4,\ 0.6,\ 0.45,\ 0.55,\ 0.5,\ 0.5)\} \\ \{(0.5,\ 1,\ 0.56,\ 1,\ 0.62,\ 1,\ 0.68,\ 1,\ 0.74,\ 1,\ 0.8,\ 1), \\ (0.5,\ 1,\ 0.56,\ 1,\ 0.62,\ 1,\ 0.68,\ 1,\ 0.74,\ 1,\ 0.8,\ 1)\}. \end{cases}$$

7.2 A New Fuzzy Nonlinear Perceptron Based on Alpha Level Sets

7.2.1 *Network Architecture*

This section proposes and analyzes a new model of Multilayer Perceptron based on interval arithmetic, which makes easier the handling of the input and output interval data, but the weights are single-valued and not interval-valued.

We suppose that we have the fuzzy rules of the form:

$\Re_p$: If x_1 is A_{p1} and x_2 is A_{p2} and ... and x_n is A_{pn} then y_1 is B_{p1} and y_2 is B_{p2} and ... and y_m is B_{pm}, $p = \overline{1, N}$, where:

- $A_{pi},\ i = \overline{1, n}$ and $B_{pj},\ j = \overline{1, m}$ are fuzzy numbers,
- N is the number of the fuzzy rules,
- n means the number of components in the antecedent side of a rule,
- m represents the number of components in the conclusion of the respective rule.

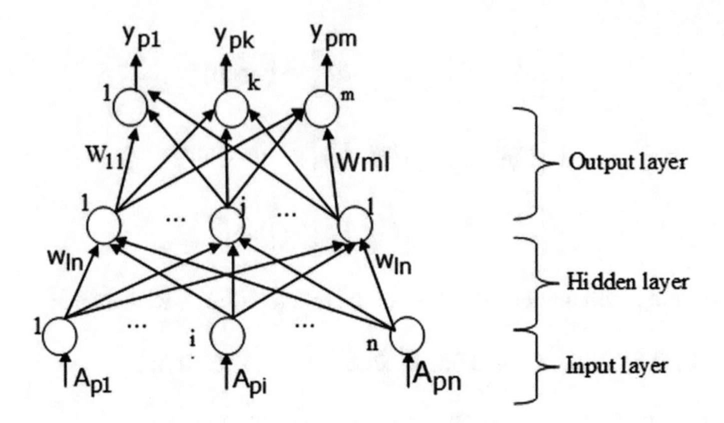

Fig. 7.6 FNPALS architecture

The following training pattern set of the network results from the previous fuzzy rules:

$$\aleph = \{(A_1, Y_1), \ldots, (A_N, Y_N)\},$$

where:

- $A_p = (A_{p1}, \ldots A_{pn})$ is the antecedent side of the p-th rule and it constitutes the input fuzzy vector of FNPALS;
- $Y_p = (Y_{p1}, \ldots Y_{pm})$ is the conclusion of the rule p and it means the fuzzy output ideal vector of our network, for each $p = \overline{1, N}$.

Figure 7.6 shows the architecture of FNPALS.
In Fig. 7.6 we have:

- $\{w_{ji}\}_{j=\overline{1,l},\ i=\overline{1,n}}$ means the set of the weights corresponding to the hidden layer,
- $\{W_{kj}\}_{k=\overline{1,m},\ j=\overline{1,l}}$ is the set of the weights corresponding to the output layer of FNPALS,
- n constitutes the number of the neurons from the input layer of FNPALS,
- l is number of the neurons in the hidden layer,
- m means the number of the neurons in the output layer, which coincides with the number of the considered classes.

The α level sets corresponding to the output of the hidden layer will be denoted by

$$\left[o_{pj}\right]^{\alpha} = \left[o_{pj}^{L}(\alpha),\ o_{pj}^{R}(\alpha)\right],\ \alpha \in [0, 1],\ j = \overline{1, l}, \tag{7.23}$$

where:

- $o_{pj}(\alpha)^L$ represents the left side of the α level sets corresponding to the outputs of the hidden layer o_{pj},
- $o_{pj}(\alpha)^R$ represents the right side of the α level sets associated to the outputs of the hidden layer o_{pj}.

We shall have:

$$\left[o_{pj}\right]^{\alpha}=\left[f\left(\sum_{i=1}^{n}w_{ji}A_{pi}\right)\right]^{\alpha}, \tag{7.24}$$

A_p being the fuzzy vector by the index p, which one applies to the network input.

Using the property that for a fuzzy number A and a continuous function f we have

$$\left[f\left(A\right)\right]^{\alpha}=f\left(\left[A\right]^{\alpha}\right), \tag{7.25}$$

the relation (7.24) becomes:

$$\left[o_{pj}\right]^{\alpha}=f\left(\sum_{i=1}^{n}\left[w_{ji}A_{pi}\right]^{\alpha}\right). \tag{7.26}$$

If f is a monotone increasing function, then (7.25) may be written in the following form:

$$\left[f\left(A\right)\right]^{\alpha}=\left[f\left(A^{L}(\alpha)\right), f\left(A^{R}(\alpha)\right)\right], \tag{7.27}$$

where

$$[A]^{\alpha}=\left[A^{L}(\alpha),\ A^{R}(\alpha)\right],\ \alpha\in[0,1]. \tag{7.28}$$

As the transfer function f of the network is a unipolar sigmoid function, i.e.,

$$f(t)=\frac{1}{1+e^{-\lambda t}},\ \lambda\in\Re, \tag{7.29}$$

therefore it is monotone increasing, then we can apply the property (7.27) to the relation (7.26) and it results:

$$\left[o_{pj}\right]^{\alpha}=\left[f\left(\sum_{i=1}^{n}(w_{ji}A_{pi})^{L}(\alpha)\right), f\left(\sum_{i=1}^{n}(w_{ji}A_{pi})^{R}(\alpha)\right)\right]. \tag{7.30}$$

Taking into account that in the right side of the relation (7.30) the weights $w_{ji},\ j=\overline{1,l},\ i=\overline{1,n}$ are not fuzzy numbers, and only the components of the input vector $A_{pi},\ p=\overline{1,N},\ i=\overline{1,n}$ are fuzzy numbers, then, using the property

$$[\lambda A]^{\alpha}=\begin{cases}\left[\lambda A^{L}(\alpha),\ \lambda A^{R}(\alpha)\right], & \text{if } \lambda\geq 0,\\ \left[\lambda A^{R}(\alpha),\ \lambda A^{L}(\alpha)\right], & \text{if } \lambda<0\end{cases} \tag{7.31}$$

the relation (7.30) may be transformed in the relation:

$$
\left[o_{pj}\right]^{\alpha} = \begin{cases} \left[f\left(\sum_{i=1}^{n} w_{ji} A_{pi}^{L}(\alpha)\right), f\left(\sum_{i=1}^{n} w_{ji} A_{pi}^{R}(\alpha)\right)\right], & \text{if } w_{ji} \geq 0, \\ \left[f\left(\sum_{i=1}^{n} w_{ji} A_{pi}^{R}(\alpha)\right), f\left(\sum_{i=1}^{n} w_{ji} A_{pi}^{L}(\alpha)\right)\right], & \text{if } w_{ji} < 0, \end{cases} \tag{7.32}
$$

for $j = \overline{1, l}$.

Analogous, the α level sets corresponding to the outputs of the output layer will be denoted by

$$
\left[O_{pk}\right]^{\alpha} = \left[O_{pk}^{L}(\alpha),\ O_{pk}^{R}(\alpha)\right],\ \alpha \in [0, 1],\ k = \overline{1, m}, \tag{7.33}
$$

where:

- $O_{pk}(\alpha)^{L}$ represents the left side of the α level sets corresponding to the outputs of the output layer O_{pk},
- $O_{pk}(\alpha)^{R}$ represents the right side of the α level sets associated to the outputs of the output layer O_{pk}.

We shall deduce:

$$
\left[O_{pk}\right]^{\alpha} = \left[f\left(\sum_{j=1}^{l} W_{kj} o_{pj}\right)\right]^{\alpha}, \tag{7.34}
$$

o_{p} being the fuzzy vector obtained at the output of the hidden layer of FNPALS, when the vector by the index p one applies to the network input.

Similar to the relation (7.26), we achieve:

$$
\left[O_{pk}\right]^{\alpha} = f\left(\sum_{j=1}^{l} \left[W_{kj} o_{pj}\right]^{\alpha}\right), \tag{7.35}
$$

namely

$$
\left[O_{pk}\right]^{\alpha} = \left[f\left(\sum_{j=1}^{l} (W_{kj} o_{pj})^{L}(\alpha)\right), f\left(\sum_{j=1}^{l} (W_{kj} o_{pj})^{R}(\alpha)\right)\right]. \tag{7.36}
$$

It is also derived from the relation (7.31):

$$
\left[O_{pk}\right]^{\alpha} = \begin{cases} \left[f\left(\sum_{j=1}^{l} W_{kj} o_{pj}^{L}(\alpha)\right), f\left(\sum_{j=1}^{l} W_{kj} o_{pi}^{R}(\alpha)\right)\right], & \text{if } W_{kj} \geq 0, \\ \left[f\left(\sum_{j=1}^{l} W_{kj} o_{pj}^{R}(\alpha)\right), f\left(\sum_{j=1}^{l} W_{kj} o_{pi}^{L}(\alpha)\right)\right], & \text{if } W_{kj} < 0, \end{cases} \tag{7.37}
$$

for $k = \overline{1, M}$.

The α level sets corresponding to the ideal outputs are denoted by

$$\left[Y_{pk}\right]^{\alpha} = \left[Y_{pk}^{L}(\alpha),\ Y_{pk}^{R}(\alpha)\right],\ \alpha \in [0, 1],\ k = \overline{1, m}. \tag{7.38}$$

For each α level set one defines a cost function, which has to be minimized; this function is

$$e_p(\alpha) = \left[e_p^L(\alpha),\ e_p^R(\alpha)\right], \tag{7.39}$$

where

$$e_p^L(\alpha) = \frac{1}{2}\sum_{k=1}^{m}\left(Y_{pk}^L(\alpha) - O_{pk}^L(\alpha)\right)^2 \tag{7.40}$$

and

$$e_p^R(\alpha) = \frac{1}{2}\sum_{k=1}^{m}\left(Y_{pk}^R(\alpha) - O_{pk}^R(\alpha)\right)^2, \tag{7.41}$$

having the following significance:

- $e_p^L(\alpha)$ represents the error between the left side of each α level set of the ideal output and the left side of each α level set of the real output;
- $e_p^R(\alpha)$ means the error between the right side of each α level set of the ideal output and the right side of each α level set of the real output.

The error function for the p-th training pattern is

$$e_p = \sum_{\alpha} e_p(\alpha), \tag{7.42}$$

with $e_p(\alpha)$ from (7.39).

7.2.2 *The Training Algorithm of FNPALS*

The training algorithm of FNPALS consists in the following steps:

Step 1 Initialize:

- the set of the weights corresponding to the hidden layer $\{w_{ji}\}_{j=\overline{1,l},\ i=\overline{1,n}}$
- the set of the weights corresponding to the output layer $\{W_{kj}\}_{k=\overline{1,m},\ j=\overline{1,l}}$ with random values.

Step 2 Repeat the Steps 3–11 for each

$$\alpha \in \{\alpha_1 \alpha_2 \ldots \alpha_{M_1}\},$$

where M_1 is the number of the α level sets considered by us and, using (7.14):

$$\alpha_j = \frac{j-1}{M_1 - 1},\ j = \overline{1, M_1}.$$

Step 3

1. Apply the fuzzy vector

$$A_p = (A_{p1}, \ldots, A_{pn})$$

to the FNPALS input, where $A_{pi},\ i = \overline{1, n}$ are some fuzzy numbers of the triangular or trapezoidal form, that represent the set of the linguistic values that the linguistic variables $x_{pi},\ i = \overline{1, n}$ take in the range specified by the linguistic terms *small*, *medium*, or *big*.
2. The membership functions of the fuzzy terms *small*, *medium*, and respective *big* are defined in the relations (6.28), (6.29), and respective (6.30).
3. The fuzzy vector A_p, which one applies to the FNPALS input one obtains from the nonfuzzy vector $X_p = (x_{p1}, \ldots, x_{pn})$ associated to the input pattern in the following way:
   - for each $x_{pi},\ i = \overline{1, n}$ we have to compute the three membership coefficients to the fuzzy sets: *small*, *medium*, and *big*;
   - we determine the greatest membership coefficient; x_{pi} is *small* or *medium* or *big* if the membership coefficient to one of the three fuzzy sets is the greatest;
   - we choose A_{pi} as that fuzzy number associated to the linguistic term that designates the respective fuzzy set.

Step 4 Compute the α level sets corresponding to the components of the fuzzy vector $A_p = (A_{p1}, \ldots, A_{pn})$:

$$\left[A_{pi}\right]^{\alpha} = \left[A_{pi}^{L}(\alpha),\ A_{pi}^{R}(\alpha)\right],\ i = \overline{1, n} \tag{7.43}$$

in function of the type of the fuzzy number A_{pi} (triangular or trapezoidal), for each $i = \overline{1, n}$.

Step 5 Establish the fuzzy output vector of the hidden layer using (7.32), where

$$\left[A_{pi}\right]^{\alpha} = \left[A_{pi}^{L}(\alpha),\ A_{pi}^{R}(\alpha)\right]$$

and

$$A_{pi}^{L}(\alpha) = \begin{cases} 0, \text{ if } A_{pi} \text{ is small} \\ 0.5 - (1-\alpha)\cdot 0.25, \text{ if } A_{pi} \text{ is medium} \\ 0.8 - (1-\alpha)\cdot 0.3, \text{ if } A_{pi} \text{ is big,} \end{cases} \quad (7.44)$$

$$A_{pi}^{R}(\alpha) = \begin{cases} 0.2 + (1-\alpha)\cdot 0.3, \text{ if } A_{pi} \text{ is small} \\ 0.5 + (1-\alpha)\cdot 0.25, \text{ if } A_{pi} \text{ is medium} \\ 1, \text{ if } A_{pi} \text{ is big,} \end{cases} \quad (7.45)$$

for all $i = \overline{1,n}$.

Step 6 Compute the fuzzy output vector from the output layer of FNPALS with the relation (7.37).

Step 7 Adjust the weights corresponding to the output layer:

$$W_{kj}(t+1) = W_{kj}(t) - \eta\alpha \cdot \frac{\partial e_p(\alpha)}{\partial W_{kj}}, \; k = \overline{1,m}, \; j = \overline{1,l}, \quad (7.46)$$

η being the learning rate.

By computing the partial derivatives

$$\frac{\partial e_p(\alpha)}{\partial W_{kj}}$$

and by substituting them in (7.46) we shall obtain:

$$\begin{aligned} W_{kj}(t+1) &= W_{kj}(t) + \\ &\eta\alpha\left(Y_{pk}^{L}(\alpha) - O_{pk}^{L}(\alpha)\right) O_{pk}^{L}(\alpha)\left(1 - O_{pk}^{L}(\alpha)\right) o_{pj}^{L}(\alpha) + \\ &\eta\alpha\left(Y_{pk}^{R}(\alpha) - O_{pk}^{R}(\alpha)\right) O_{pk}^{R}(\alpha)\left(1 - O_{pk}^{R}(\alpha)\right) o_{pj}^{R}(\alpha), \end{aligned} \quad (7.47)$$

for $k = \overline{1,m}, \; j = \overline{1,l}$.

Step 8 Adjust the weights corresponding to the hidden layer:

$$w_{ji}(t+1) = w_{ji}(t) - \eta\alpha \cdot \frac{\partial e_p(\alpha)}{\partial w_{kj}}, \; j = \overline{1,l}, \; i = \overline{1,n}, \quad (7.48)$$

namely

$$w_{ji}(t+1) = w_{ji}(t) + \eta\alpha\cdot$$

$$\sum_{k=1}^{m} \left(Y_{pk}^{L}(\alpha) - O_{pk}^{L}(\alpha)\right) O_{pk}^{L}(\alpha) \left(1 - O_{pk}^{L}(\alpha)\right) W_{kj} o_{pj}^{L}(\alpha) \left(1 - o_{pj}^{L}(\alpha)\right) A_{pi}^{L}(\alpha) +$$

$$\eta\alpha \sum_{k=1}^{m} \left(Y_{pk}^{R}(\alpha) - O_{pk}^{R}(\alpha)\right) O_{pk}^{R}(\alpha) \left(1 - O_{pk}^{R}(\alpha)\right) W_{kj} o_{pj}^{R}(\alpha) \left(1 - o_{pj}^{R}(\alpha)\right) A_{pi}^{R}(\alpha), \tag{7.49}$$

for $j = \overline{1, n},\ i = \overline{1, l}$.

Step 9 Compute the fuzzy error $e_p(\alpha)$ corresponding to the training vector p using (7.39) and then the error function because of the vector p with (7.42).

Step 10 If $p < N$ (N is the number of the vectors from the training lot), we have to increment the value of p in the following way:
$p = p + 1$, namely go to the next vector of the training set and take back the algorithm from the *Step 3*.
Otherwise, go to the *Step 2*.

Step 11 Compute the error of this epoch:

$$E = \frac{1}{N} \sum_{p=1}^{N} e_p. \tag{7.50}$$

Step 12 Go to the next training epoch and repeat the algorithm from the *Step 2*.

In [2] we also used the two types of databases: portraits (belonging to the ORL Database of Faces, represented in Fig. 2.19) and an ECG database (of 40 subjects: 20 patients of Ischemic Heart Disease (IHD) depicted in Fig. 7.5 and other 20 normal subjects from Fig. 7.4) in order to test the performances of our FNPALS. With FNPALS, we obtain the best recognition score of: 96% for face recognition, 95% for ECG signals classification, while using a MLP or FNP we obtain a lower recognition score.

References

1. I. Iatan. *Neuro-Fuzzy Systems for Pattern Recognition (in Romanian)*. PhD thesis, Faculty of Electronics, Telecommunications and Information Technology- University Politehnica of Bucharest, PhD supervisor: Prof. dr. Victor Neagoe, 2003.
2. I. Iatan and M. de Rijke. A new interval arithmetic based neural network. (work in progress), 2014.
3. A. Muñoz San Roque, C. Maté, J. Arroyo, and A. Sarabia. iMLP: Applying multi-layer perceptrons to interval-valued data. *Neural Processing Letters*, 25:157–169, 2007.
4. C. T. Leondes. *Fuzzy Logic and Expert Systems Applications*. San Diego, Academic Press, 1998.
5. M. Umano and Y. Ezawa. Execution of approximate reasoning by neural network. In *Proceedings of FAN Symposium*, pages 267–273, 1991.
6. K. Uehara and M. Fujise. Learning of fuzzy inference criteria with artificial neural network. In *Proc. 1st Int. Conf. on Fuzzy Logic and Neural Networks*, pages 193–198, 1990.

Chapter 8
A Recurrent Neural Fuzzy Network

Besides the feedforward neural networks, there are the recurrent networks, where the impulses can be transmitted in both directions due to some reaction connections in these networks. Recurrent Neural Networks (RNNs) are linear or nonlinear dynamic systems. The dynamic behavior presented by the recurrent neural networks can be described both in continuous time, by differential equations and at discrete times by the recurrence relations (difference equations). The distinction between recurrent (or dynamic) neural networks and static neural networks is due to recurrent connections both between the layers of neurons of these networks and within the same layer, too. The aim of this chapter is to describe a Recurrent Fuzzy Neural Network (RFNN) model, whose learning algorithm is based on the Improved Particle Swarm Optimization (IPSO) method.

In general, there are [1] some classical algorithms for training the RNNs, proving that the RNNs are harder to learn than the feedforward neural networks, such that: Backpropagation Through Time (BPTT), Real-Time Recurrent Learning (RTRL), Extended Kalman Filter (EKF).

In the case of the IPSO method, each particle (candidate solution), which is moving permanently includes the parameters of the membership function and the weights of the recurrent neural-fuzzy network; initially, their values are randomly generated. The RFNN presented in the paper [2] is unlike the others variants of RFNN models, by the number of the evolution directions that they use: in the paper, we update the velocity and the position of all particles along three dimensions, while in [3] are used two dimensions.

8.1 Introduction

Neural network (NN) is one of the important components in Artificial Intelligence (AI). NN architectures used in modeling of the nervous systems can be classified into

I.F. Iatan, *Issues in the Use of Neural Networks in Information Retrieval*,
Studies in Computational Intelligence 661, DOI 10.1007/978-3-319-43871-9_8

three categories, each with a different philosophy: feedforward, recurrent (feedback), self-organizing map. Neural networks (NNs) are used in many different application domains in order to solve various information processing problems. For several years now, neural network models have enjoyed wide popularity [4], being applied to problems of regression, classification, computational science, computer vision, data processing and time series analysis.

The main drawback of the feedforward neural networks is that the updating of the weights can fall [5] in a local minimum. Another major drawback of the feedforward neural networks consists in the fact that their application domain is limited to static problems due their inherent feedforward structure.

Since recurrent networks incorporate feedback, they have powerful representation capability and can [5] successfully overcome disadvantages of feedforward networks. This feedback implies that the network has [6] local memory characteristics that is able to store activity patterns and present those patterns to the network more than once, allowing the layer with feedback connections to use its own past activation in its preceding behavior.

The neurons of the Recurrent Neural Networks (RNNs) are combined in a possibly cyclic graph such that [7]:

- to notice the time dependent dynamics;
- the outputs corresponding to all the neurons are computed on the activations in the previous time step.

The RNN has the feedforward and feedback connections contrasted which provides it with nonlinear mapping capacity and dynamical characteristics, so it can be used [8] to simulate dynamical system and solve dynamic problems. Different architectures can be created [6] by adding recurrent connections at different points in the basic feedforward architecture, like:

- Hopfield networks;
- Bidirectional Associative Memory (BAM), a generalization of the Hopfield network;
- Carpenter–Grossberg networks, that are based on the *Adaptive Resonance Theory* (ART);
- Jordan recurrent neural networks (JNNs);
- Wavelet-based neuro-fuzzy networks (WNFNs);
- Selective recurrent neural networks (SRNNs).

Recently some researchers have proposed several recurrent neuro-fuzzy networks. Kumar et al. (2004) compares the traditional feedforward approach of RNNs to forecast monthly river flows. Lin and Hsu (2007) has proposed [17] a recurrent wavelet-based neuro-fuzzy system with the reinforcement hybrid evolutionary learning algorithm for solving various control problems. Carcano et al. (2008) has simulated [10] daily river flows for water resource purposes using the Jordan Recurrent Neural Network. Maraqua et al. (2012) has proposed [6] the use of a recurrent network architecture as a classification engine for automatic Arabic Sign Language recognition system. Šter (2013) has introduced [18] an extended architecture of recurrent neural networks (called *Selective Recurrent Neural Network*) for dealing with long term dependencies.

8.1.1 Wavelet Neural Networks

Neural networks employing wavelet neurons are referred to as Wavelet Neural Networks (WNNs) [17]; they are characterized by weights and wavelet bases.

A fuzzy wavelet neural network (FWNN) has the aim [11] to combine the wavelet theory with the fuzzy logic and the NNs.

Lin and Chin (2004) was proposed a Recurrent Neural Fuzzy Network (RNFN) where each fuzzy rule corresponding to a WNN (see Fig. 8.1) consists (see [3, 12]) of single-scaling wavelets. The shape and position of the wavelet bases are shown [12] in Fig. 8.2.

An ordinary wavelet neural network model is often used to normalize input vectors in the interval [0, 1]. The functions $\phi_{a.b}(x_i)$ are used to input vectors to fire up the

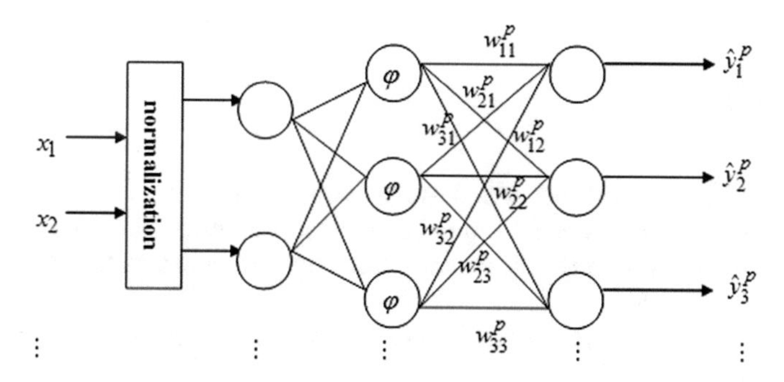

Fig. 8.1 Schematic diagram of the WNN

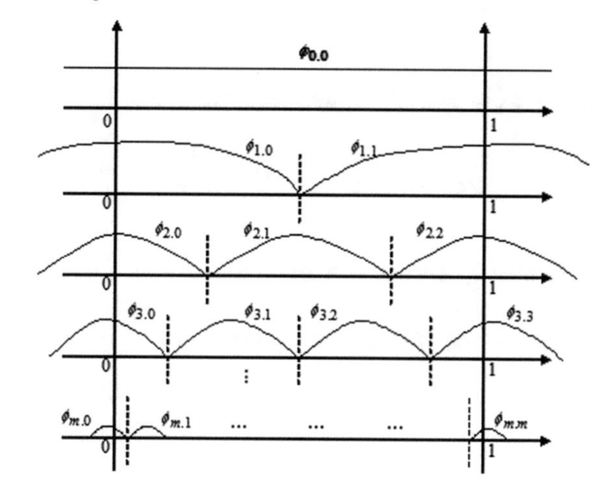

Fig. 8.2 Wavelet bases are over-complete and compactly supported

wavelet interval; a such value is given in the following equation, which gives the shape of the M wavelet bases $\phi_{0.0},\ \phi_{1.0},\ldots,\phi_{m.m}$:

$$\begin{cases} \phi(x_i) = \cos(x_i),\ -0.5 \leq x_i \leq 0.5 \\ 0 \text{ otherwise}, \quad \phi_{a.b}(x_i) = \cos(ax_i - b), \end{cases} \tag{8.1}$$

$b = \overline{1, a},\ a = \overline{1, m}$, b being a shifting parameter and a meaning a scaling parameter corresponding to the maximum value of b.

A crisp value $\varphi_{a.b}$ can be obtained as follows:

$$\varphi_{a.b} = \frac{\sum_{j=1}^{n} \phi_{a.b}(x_i)}{|X|}, \tag{8.2}$$

where $|X|$ represents the number of input dimensions and n is the dimension of the input vector to the model.

8.1.2 Z-Transform

The Z-transform is [14] the discrete-time counterpart of the Laplace transform. The Z-transform can be considered to be an extension of the discrete-time Fourier transform as the Laplace transform can be considered an extension of the Fourier transform.

The *bilateral* Z-transform of a discrete-time sequence $x(n)$ is:

$$Z\{x(n)\} = X(z) = \sum_{n=-\infty}^{\infty} x(n)z^{-n}. \tag{8.3}$$

For causal sequences ($n \geq 0$) the Z-transform becomes:

$$Z\{x(n)\} = X(z) = \sum_{n=0}^{\infty} x(n)z^{-n}. \tag{8.4}$$

The Eq. (8.4) is called the *unilateral* Z-transform; it exists only if the power series from its expression converges.

There are several methods for computing the inverse Z-transform, namely the sequence $x(n)$, given $X(z)$:

1. using the *inversion integral*

$$x(n) = \frac{1}{2\pi j} \oint_{\Gamma} X(z) z^{n-1} dz, \tag{8.5}$$

   where $\oint_{\Gamma}$ means the integration along the closed contour Γ in the counterclockwise closed contour in the region of convergence of $X(z)$;
2. by a power series expansion: expressing $X(z)$ in a power series in z^{-1}, $x(n)$ can be achieved by identifying it with the coefficient of z^{-n} in the power series expansion;
3. by partial fraction expansion: for a rational functions, can be obtained a partial fraction expansion of $X(z)$ over its poles and the table of Z-transform helps to identify the sequences corresponding to the terms in that partial fraction expansion.

8.1.3 Application of Genetic Algorithms

The specialists think that the Genetic Algorithms are a computational intelligence application as well as the expert systems, fuzzy systems, neural networks, the intelligent agents, hybrid intelligent systems, electronic voice.

The genetic algorithms are some adaptive techniques of heuristic search, based on the genetic and selection natural principles, enunciated by Darwin (the best adapted will survive). The mechanism is similar to the evolutionary biological process. This process has a feature through that only the species which one adapt better to the environment are capable to survive and to develop into generations, while that those less adapted fail to survive and they disappear in time, as a result of the natural selection. The main notions that allow the analogy between the solution of the search problems and the natural evolution are [15]:

1. *Population*. A population consists in some individuals (*chromosomes*) that have to live in an environment to which they must adapt.
2. *Fitness*. Each of the population individuals is adapted more or less to the environment. The fitness is a measure of the degree of adaptation to the environment.
3. *Chromosome*. It is an ordered set of elements, named *genes*, whose values establish the individual features.
4. *Generation*. A stage in a population evolution. If we see the evolution as an iterative process in which a population turns to another population, then the generation is an iteration in this process.
5. *Selection*. The process of natural selection has the survival of individuals with a high environmental fitness (high fitness) as effect.
6. *Reproduction*. It is the process through which one passes from one generation to another. The individuals of the new generation inherit some features from their precursors (parents) but they can also get some new features as a result of

some processes of mutation that have a random character. In the case when in the reproduction process at least two parents occur, the inherited features of the survivor (son) are obtained by combining (crossover) of the parent features.

The remainder of the chapter is organized as follows. In Sect. 8.2 is discussed and analyzed the RNFN. We follow with the learning algorithm of the recurrent model in Sect. 8.3.

8.2 RNFN Architecture

The network construction is based on fuzzy rules, each corresponding to a Wavelet Neural Network (WNN).

The Fig. 8.3 illustrates the RNFN model, whose training algorithm is based on Improved Particle Swarm Optimization (IPSO) method.

The nodes from the first layer constitute some input nodes; hence they only pass the input signal to the next layer, namely [3]:

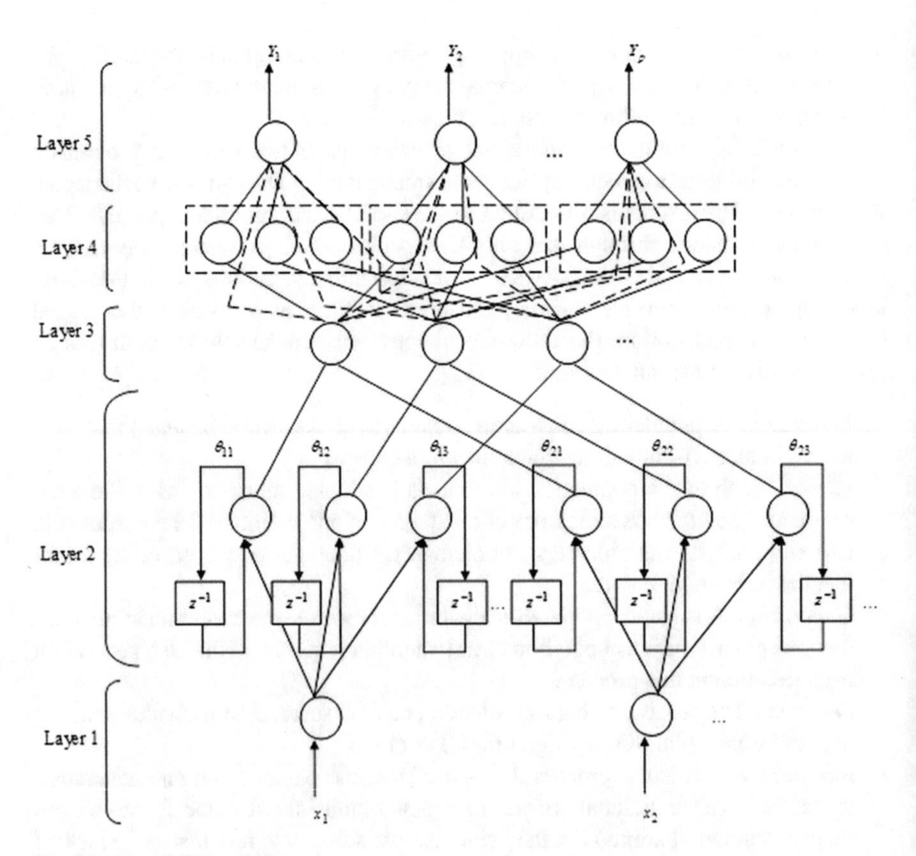

Fig. 8.3 The RNFN architecture

$$O_i^{(1)} = x_i^{(1)}. \tag{8.6}$$

The neurons in the second layer act as a membership function, meaning that they determine how an input value belongs to a fuzzy set. The following Gaussian function is chosen as the membership function [3]:

$$O_{ij}^{(2)} = \mathrm{e}^{-\frac{\left(I_{ij}^{(2)} - m_{ij}\right)^2}{\sigma_{ij}^2}}, \tag{8.7}$$

where:

- m_{ij} and σ_{ij} are the mean and standard deviation, respectively;
- $I_{ij}^{(2)}$ denotes the input of this layer for the discrete time scan:

$$I_{ij}^{(2)} = O_i^{(2)} + O_{ij}^{(f)}, \tag{8.8}$$

where

$$O_{ij}^{(f)} = O_{ij}^{(2)}(t-1)\theta_{ij}. \tag{8.9}$$

The inputs of this layer contain the terms of memory $O_{ij}^{(2)}(t-1)$, that store network information at a previous time; this information, which is an additional input of the network will be reintroduced at the entrance of the second layer.

The weight θ_{ij} constitutes the feedback weight of the network and z^{-1} signifies the delayed operator.

Figure 8.4 represents [16] a delayed cell, $X(z)$ being the Z-transform of the signal $x[n]$.

The neurons of the third layer achieve the product operation of their input signals [3]:

$$O_j^3 = \prod_{i=1}^{n} O_{ij}^{(2)} = \prod_{i=1}^{n} \mathrm{e}^{-\frac{\left(I_{ij}^{(2)} - m_{ij}\right)^2}{\sigma_{ij}^2}}, \tag{8.10}$$

where n is the number of external dimensions.

The neurons of the fourth layer receive both the output of a WNN, denoted $\hat{y}_j$ and of a neuron from the third layer, namely O_j^3. The mathematical function of each node j is [3]:

$$O_j^4 = \hat{y}_j^p \cdot O_j^3, \tag{8.11}$$

$\hat{y}_j^p$ being the local output of the WNN for the output y_p and the j-th rule:

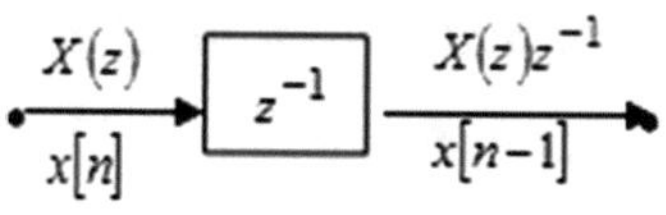

Fig. 8.4 Delayed cell

$$\hat{y}_j^p = \sum_{k=1}^{M} w_{jk}^p \varphi_{a.b}, \tag{8.12}$$

with $\varphi_{a.b}$ from (8.2), where:

- $M = m + 1$ denotes the number of wavelet bases, which equals the number of existing fuzzy rules in the considered model,
- the link w_{jk}^p is the output action strength associated in the p output, with j-th rule and k-th $\varphi_{a.b}$.

The fifth layer acts as a defuzzifier, namely it provides the nonfuzzy outputs y_p of the fuzzy recurrent neural network [3]:

$$y_p = \frac{1}{1 + \mathrm{e}^{-\lambda \cdot \frac{\sum_{j=1}^{M} o_j^4}{\sum_{j=1}^{M} o_j^3}}} = \frac{1}{1 + \mathrm{e}^{-\lambda \cdot \frac{\sum_{j=1}^{M} \hat{y}_j^p \cdot o_j^3}{\sum_{j=1}^{M} o_j^3}}}, \tag{8.13}$$

namely

$$y_p = \frac{1}{1 + \mathrm{e}^{-\lambda \cdot \frac{\sum_{j=1}^{M} (w_{j1}^p \varphi_{1.1} + w_{j2}^p \varphi_{2.1} + \cdots + w_{jM}^p \varphi_{m.m}) \cdot o_j^3}{\sum_{j=1}^{M} o_j^3}}}, \quad \lambda \in \Re. \tag{8.14}$$

8.3 Learning Algorithm of RNFN

The training algorithm of the network is based on the Improved optimization method Particle Swarm Optimization (IPSO). The new optimization algorithm called the IPSO enhances the traditional PSO (Particle Swarm Optimization) to enable it to obtain optimal solution capability. We assume that each particle includes the mean, deviation and weight variables of the RNFN, being d-dimensional.

The following parameters will be determined by the learning procedure [3]:

- the position vector $X_i = (x_{i1}, x_{i2}, \ldots, x_{id})$,
  and, respectively
- the velocity vector $V_i = (v_{i1}, v_{i2}, \ldots, v_{id})$

of the i-th particle in the N-dimensional search space.

We denote by:

- $P_i = (P_{i1}, P_{i2}, \ldots, P_{id})$ the best position of each particle,
- $P_g = (P_{g1}, P_{g2}, \ldots, P_{gd})$ the fittest particle found so far,

according to an user-defined fitness function.

The steps of the learning procedure are [3]:

Step 1 (*Individual initialization*). Set the initial values for every particle like being random values.

Step 2 (*Evaluate fitness*). Evaluate each particle in a swarm, by defining the fitness function:

$$f_i = \frac{1}{Y}, \tag{8.15}$$

where:

$$Y = \sqrt{\frac{1}{N}\sum_{p=1}^{N}(y_p - \overline{y}_p)^2}, \tag{8.16}$$

- N represents the number of input data,
- $y_p,\ p = \overline{1, N}$ are the model outputs,
- $\overline{y}_p,\ p = \overline{1, N}$ constitute the desired outputs.

After a generation of learning, we achieve the following fifth best particles, ordered according to their fitness: *unimportant*, *rather unimportant*, *moderately important*, *rather important*, *very important* particles.
The input (preferred) particles are:

1. *unimportant* particle

$$C_u = (C_{u1}, C_{u2}, \ldots, C_{ud}),$$

with the fitness F_u;

2. *rather unimportant* particle

$$C_r = (C_{r1}, C_{r2}, \ldots, C_{rd}),$$

with the fitness F_r;

3. *moderately important* particle

$$C_m = (C_{m1}, C_{m2}, \ldots, C_{md}),$$

with the fitness F_m;

4. *rather important* particle

$$C_R = (C_{R1}, C_{R2}, \ldots, C_{Rd}),$$

with the fitness F_R;

5. *very important* particle

$$C_v = (C_{v1}, C_{v2}, \ldots, C_{vd}),$$

with the fitness F_v.

The membership functions of the fuzzy terms *unimportant*, *rather unimportant*, *moderately important*, *rather important*, and respectively *very important* can be represented as fuzzy numbers in Fig. 1.13, being defined in the following relations (1.14)–(1.18).
The output (created) particle is *output particle*

$$C_o = (C_{o1}, C_{o2}, \ldots, C_{od}),$$

with the fitness F_o.

Step 3 (*Improve the capability of finding the global solution* (ICFGS)). Set: $D_1 = D_2 = D_3 = 1$ the magnitudes of the three evolution directions, $T_s = 1$ the initial index of the ICFGS, the number N_L of the ICFGS loop, the fifth particles with the best fitness values from the local best swarm to C_u, C_r, C_m, C_R, C_v.
Use a special equation to update the *unimportant* particle, *rather unimportant* particle, *moderately important* particle and *rather important* particle to generate the migrant individuals, based on the best individual, $X_i = (x_{i1}, \ldots, x_{id})$ in the aim of improving the fitness value [13]:

$$x_{id} = \begin{cases} x_{id} + \rho(x_{id}^L - x_{id}), \text{ if } r_1 < \frac{x_{id} - x_{id}^L}{x_{id}^L - x_{id}^U} \\ x_{id} + \rho(x_{id}^U - x_{id}) \text{ otherwise,} \end{cases} \tag{8.17}$$

where ρ and r_1 are random numbers in the range of [0, 1] and L, U meaning "lower" and "upper".

Compute C_o:

$$C_{oj} = C_{uj} + D_1(C_{uj} - C_{rj}) + \\ + D_2(C_{uj} - C_{mj}) + D_3(C_{uj} - C_{Rj}) \tag{8.18}$$

Evaluate the new fitness F_o corresponding to the newly created output particle C_o.
Update the *unimportant* particle C_u, the *rather unimportant* particle C_r, *moderately important* particle C_m, *rather important* particle C_R and the *very important* particle C_v as follows:

(1) If $F_o > F_v$ then:

$$\begin{cases} C_v = C_o \\ C_R = C_v \\ C_m = C_R \\ C_r = C_m \\ C_u = C_r. \end{cases}$$

(2) Else if $F_o > F_R$ and $F_o < F_v$ then:

$$\begin{cases} C_R = C_o \\ C_m = C_R \\ C_r = C_m \\ C_u = C_r. \end{cases}$$

(3) Else if $F_o > F_m$ and $F_o < F_R$ then:

$$\begin{cases} C_m = C_o \\ C_r = C_m \\ C_u = C_r. \end{cases}$$

(4) Else if $F_o > F_r$ and $F_o < F_m$ then:

$$\begin{cases} C_r = C_o \\ C_u = C_r. \end{cases}$$

(5) Else if $F_o > F_u$ and $F_o < F_r$ then:

$$C_u = C_o.$$

(6) Else if $F_o = F_u = F_r = F_m = F_R = F_v$ then:

$$C_o = C_o + N_r \; (N_r \in [0, 1]). \tag{8.19}$$

(7) Else if $F_o <= F_u$ then it will decrease the moving velocity:

$$\begin{cases} D_1 = -0.5D_1 \\ D_2 = -0.5D_2 \\ D_3 = -0.5D_3 \end{cases}$$

to obtain a good fitness.
The random number N_r is added at the statement (8.19) to prevent the learning algorithm from falling into a local optimum.
Test If *Step 3* is not finished then $T_s = T_s + 1$; else update the global best: if the fitness value of the new particle is higher than that of the global best, then the global best will also be replaced with the particle.

Step 4 (*Update the velocity and the position*). Update the velocity and the position of all particles along each dimension using the equations:

$$v_{id}^{k+1} = \omega \cdot v_{id}^{k} + c_1 \cdot \text{rand}(\cdot)(P_{id} - x_{id}^{k}) + c_2 \cdot \text{rand}(\cdot)(P_{gd} - x_{id}^{k}) \tag{8.20}$$

$$x_{id}^{k+1} = x_{id}^{k} + v_{id}^{k+1}, \tag{8.21}$$

where w is the coefficient of the inertia term; c_1 and c_2 are called the cognitive term and the society term, respectively; the function rand$(\cdot)$ yields uniformly distributed random numbers in [0, 1].

The second term from (8.20) known as the cognitive component, represents the personal thinking of each particle, which encourages the particles to move toward their own best positions. The third term from (8.20) called the social component represents the collaborative effect of the particles, in finding the global optimal solution.

Dynamic or Recurrent Neural Networks (RNNs) are unlike from static neural networks since they include feedback or recurrent connections between the network layers and within the layer itself.

The learning algorithm of the Recurrent Neural Fuzzy Network (RNFN) model presented in this chapter is based on the Improved Particle Swarm Optimization (IPSO) method, which is similar to evolutionary algorithms, but requires less computational bookkeeping and generally fewer lines of code. The new optimization algorithm called the IPSO enhances the traditional PSO (Particle Swarm Optimization) to enable it to obtain optimal solution capability.

The RFNN presented in this chapter is unlike the others variants of RFNN models, by the number of the evolution directions that they use: in this chapter, we update the velocity and the position of all particles along three dimensions.

The network construction is based on fuzzy rules, each corresponding to a WNN (Wavelet Neural Network).

References

1. G. Strong and M. Gong. Similarity-based image organization and browsing using multi-resolution self-organizing map. *Image and Vision Computing*, 29:774–786, 2011.
2. G.A. Anastassiou and I. Iatan. A recurrent neural fuzzy network. *Journal of Computational Analysis and Applications*, 20 (2), 2016.
3. C.J. Lin, M. Wang, and C.Y. Lee. Pattern recognition using neural-fuzzy networks based on improved particle swam optimization. *Expert Systems with Applications*, 36:5402–5410, 2009.
4. J. Golbeck, C. Robles, and K. Turner. Predicting personality with social media. In *Proceedings of alt.chi, ACM Conference on Human Factors in Computing*, pages 253–262, 2012.
5. Rubio J.J. Stability analysis for an online evolving neuro-fuzzy recurrent network. In *Evolving Intelligent Systems: Methodology and Applications*, pages 173–199. Wiley-IEEE Press, 2010.
6. M. Maraqua, F. Al-Zboun, M. Dhyabat, and R.A. Zitar. Recognition of arabic sign language (ArSL) using recurrent neural networks. *Journal of Intelligent Learning Systems and Applications*, 4:41–52, 2012.

7. B. Hammer and T. Villmann. Mathematical aspects of neural networks. In *11th European Symposium on Artificial Neural Networks (ESANN' 2003)*, 2003.
8. Zhao F., Hu L., and Li Z. Nonlinear system identification based on recurrent wavelet neural network. In *Advances in Intelligent and Soft Computing*, volume 56, pages 517–525. Springer, 2009.
9. X. Li, C. G. M. Snoek, and M. Worring. Learning tag relevance by neighbor voting for social image retrieval. In *Proceedings of the 1st ACM international conference on Multimedia information retrieval*, pages 180–187, 2008.
10. Carcano E.C., Bartolini P., Muselli M., and Piroddi L. Jordan recurrent neural network versus ihacres in modelling daily streamflows. *Journal of Hydrology*, 362:291–307, 2008.
11. R.H. Abiyev and O. Kaynak. Identification and control of dynamic plants using fuzzy wavelet neural networks. In *Proceedings of the IEEE Multi-conference on Systems and Control*, pages 1295–1301, 2008.
12. Lin C.J. and Chin C.C. Recurrent wavelet-based neuro fuzzy networks for dynamic system identification. *Mathematical and Computer Modelling*, 41:227–239, 2005.
13. W. Lin, D. Tao, J. Kacprzyk, Z. Li, E. Izquierdo, and H. Wang. *Multimedia Analysis, Processing and Communications*. Springer-Verlag Berlin Heidelberg, 2011.
14. R.A. Tuduce. *Signal Theory*. Bren, Bucharest, 1998.
15. D. Dumitrescu. *Genetic algorithms and Evolution strategies- applications in Artificial Intelligence and in conex domains (in Romanian)*. Blue Publishing House, Cluj-Napoca, 2006.
16. A. Mateescu. *Traitement numérique des signaux*. Editions Techniques, Bucarest, 1997.
17. Lin C.J., Hsu Y.C.: Reinforcement hybrid evolutionary learning for re-current wavelet-based neuro-fuzzy systems. *IEEE Transactions on Fuzzy Systems*, 15(4):729–745 (2007).
18. Ster B.: Selective recurrent neural network. *Neural Processing Letters*, 38:1–15 (2013).

Printed in the United States
By Bookmasters